UNDERSTANDING

ELEVATOR TECHNOLOGY

WITH

FOCUS ON ELECTRIC TRACTION LIFTS

BY

PRASAD DANDAPANI

ISBN 9781648051647

PREFACE

Owing to the steep increase in growth of population, the cities have started growing vertically resulting in large number of high rise buildings for office, commercial and residential spaces. Vertical transportation of people has now become a necessity just as the surface and air transportation. In addition to providing comfort and safety to the passengers, the elevators also help to add to the beauty of the building. Number of elevators with proper capacity and speed improve the passenger handling thus saving travelers' valuable time in waiting and transit.

The architects and builders decide the quantity, capacity, speed and other specifications of the elevators within the budget constraints. But the actual users of the building are the residents, who own the premises once the building is completed and handed over. Since the architects and builders become accountable for the actual users, they take the help of elevator consultants while deciding the elevators for the building.

This book aims at providing basic technical information to the builders and architects which they normally seek from the consultants.

Lot of work has been done by experts in the field of elevators and many articles have been written on this subject though there are very few books on this subject. This book is basically a compilation of material in the books, magazines and the net which the author came across combined with the author's knowledge gained by experience working in elevator industries for over 30 years.

The content of this book is expected to provide basic elevator knowledge to the students, particularly future Engineers, builders and architects. It is also my aim to help the employees of elevator companies, to get to know elevator product in totality.

Currently students do not get the opportunity to study about elevators. This book could lay the foundation for introducing " Basics of Elevator technology" as an elective subject. It is the author's belief that engineers who have done an elective in elevator engineering would find it easy to get absorbed in the Elevator industry. This subject can be taken as an elective by Civil, Mechanical, Electrical and Electronics branch of engineering and also by architects.

This book may also be a source of knowledge for the common man who manages housing societies.

FOREWORD-1

I have known Mr D Prasad for many years as an able Engineering professional. As our Engineering Head Mr.Prasad participated in many world-wide engineering meetings and has visited many elevator factories and OTIS engineering centers across the world. For systems integration, he was also reporting to our worldwide Engineering leadership of our company.

Though Mr Prasad was well known to me as a dedicated engineering professional, I was not aware about his passion for converting the essence of his learning's into a book form. He has done a commendable job in his first attempt at writing. This book seems to have been written primarily for engineers, architects and students who want to get up to speed on basics of elevator technology. Rather than assuming any particular level of proficiency, concepts are explained from the point of view of a novice.

This book will help cover the vacuum on elevator literature for their design basics and qualitative application details including traffic analysis. I believe that this book can ignite curiosity and excitement to get more knowledge in this subject. I recommend this book to engineers seeking information on elevator engineering.

Ashok Malhotra
Former Managing Director
OTIS Elevator Co India Ltd

FOREWORD – 2

There are some things we interact with every day without ever understanding how they work. Elevators are one of them. Behind the simple action of pushing a button, stepping into a cabin and getting to an upper or a lower floor of a building, there is technology but there is most of all the passion of professionals like Prasad who have dedicated part of their career in designing elevators. Therefore, I appreciate seeing him taking the time to write a book that gets to the essence of elevators in very simple terms for engineers, architects, students, general managers, and any one with the curiosity to go behind the scenes to understand how they work and the governing rules that drive their specification and design.

Leandre Adifon

**Former Vice President of
OTIS Worldwide Engineering, USA**

DISCLAIMER:

AT Some Parts in this Book RELEVANT EXTRACTS OF INDIAN CODES HAVE BEEN REFERRED FOR EASY UNDERSTANDING BY THE READER. HOWEVER IT SHOULD NOT BE CONSTRUED THAT THE CODES GIVEN IN THIS BOOK ARE COMPLETE AND ADEQUATE FOR CARRYING OUT ACTUAL WORK . THERE COULD BE REVISIONS TO THE CODES ALSO. ARCHITECTS, BUILDERS AND OTHER USERS ARE STRONGLY ADVISED TO ABIDE BY ALL THE LATEST VERSIONS OF CODES AND OTHER REGULATIONS AS PUBLISHED BY GOVERNMENT OF INDIA AND OTHER AUTHORITIES. THIS BOOK ONLY PROVIDES A OVERVIEW OF OPERATION OF ELEVATORS.

ABOUT THE AUTHOR

Mr Prasad Dandapani is a professional in the field of elevator technology with about four decades of experience. He is a graduate in Electronics and Communication Engineering from the College of Engineering, Guindy, Madras University and MBA from the Indira Gandhi National Open University.

He started his career with a leading elevator company in India ECE Elevators and later moved to OTIS, a global leader in elevators. In OTIS India, he served as the head of engineering department and has vast international exposure in engineering design. He has visited many manufacturing units and engineering centers internationally. He is also well trained in quality management systems. After retirement he continues his contribution to M/S Johnson Lifts

Private Limited, India's leaders in elevator industry as technical advisor.

This book is a condensation of the vast and profound experience in the Elevator Industry and compilation from books, journals and the internet. This book is expected to serve as a valuable guide for everyone connected with lift industry and those planning to enter this field.

CONTENTS

CHAPTER 1

Elevator – Basic Concept

1.1 What is an Elevator or Lift?

Elevators or Lifts are used for vertical transportation. We can also define elevator as "A hoisting and lowering mechanism equipped with a car or platform which moves along the guides in a vertical direction and which serves two or more floors of a building".

1.2 Why do we need Elevators?

Elevators have become a necessity by virtue of the human comfort and convenience, and or by statutory regulations. The increase in the number of high and medium rise buildings in the last five decades has given rise to large number of elevators. During the last decade, large improvements have taken place in the engineering of elevator systems with introduction of superior drives, logic control systems and elegant cars with enhanced safety.

1.3 Invention of Elevator:

Crude elevators with steam or hydraulics as source of power were available during 1800s. The first hydraulic elevators were designed using water pressure as the source of power. In 1853 Elisha Graves Otis created the first safety hoist. He installed a rope-break safety device called the safety brake (the equivalent of the modern safety gear) into the elevator. With the Otis safety brake, in the case of rope failure, a spring would force a ratchet to engage saw-tooth iron bars and safely secure the car from falling. In 1854, Otis demonstrated the safety brake by boarding his elevator at the Crystal Palace in New York and cutting the traditional hemp rope himself. The safety brake worked flawlessly,

13

making a dramatic presentation. "All Safe, gentlemen", Elisha Otis announced as the brakes kicked in. By executing this stunt,, Otis had heralded the birth of the elevator industry. Otis established a company for manufacturing elevators and went on to dominate the elevator industry. Today the Otis Elevator Factory is one of the world's largest manufacturers of vertical transport systems.

1.4 When was the first electric powered elevator displayed?

At the Mannheim Exposition of 1880, as the industrialized world was adopting electrical power, the German company Siemens exhibited an electric powered elevator. The mid-19th century marked the dawning of the age of electricity, and developments in elevator technology was being driven by the appearance of the first high-rise buildings in the United States, which necessitated the development of elevators in order to make them practical. As such, the United States emerged as the center of elevator technology development for decades.

1.5 What is the normal speed of elevators installed in India?

Popular speed of elevators in India are from 0.63 meters per second (2.2 KM/H) to 2.5 meters per second (9 KM/H).

1.6 What is the fastest speed of elevator?

As per the understanding of the author, the fastest speed is 72KM/H (20mps)by Mitsubishi lifts for Shanghai tower, China. Burj Khalifa in Dubai has Otis elevators traveling at 36 KM/H.(10mps)

1.7How does the elevator function?

To learn about the function of elevators, we shall consider the analogy of fetching water from a well. In Indian villages, many would have observed how the water is fetched from a well. A bucket is tied to a rope and is passed through a pulley. The bucket tied to the rope is lowered using the pulley and once it reaches the bottom, water gets filled inside the bucket and then the bucket is pulled up manually.

From the above analogy, it is easy to develop the concept of operation of elevators. First replace the bucket with a lift cage, rectangular in shape, which will be used to carry the passengers. This lift cage is called a **car**, in elevator terminology. If the rectangular lift car has to move up and down the well, the well also has to be rectangular in shape. This rectangular shape of well is known as elevator shaft or Hoistway.

1.8 Car & Counterweight:

In our above analogy, we had replaced the bucket with rectangular elevator car. In elevators, one side of the rope is tied to the car and the other side of the rope is tied to a counter weight. The counter weight consists of a counter weight frame filled with cast iron weights. Nowadays Concrete filler weights are also used primarily for reducing the cost.

A counterweight is an equivalent counterbalancing weight that balances the car and its passenger load. Typically, the counterweight is the weight of the car plus 40-50% of the car's rated capacity.

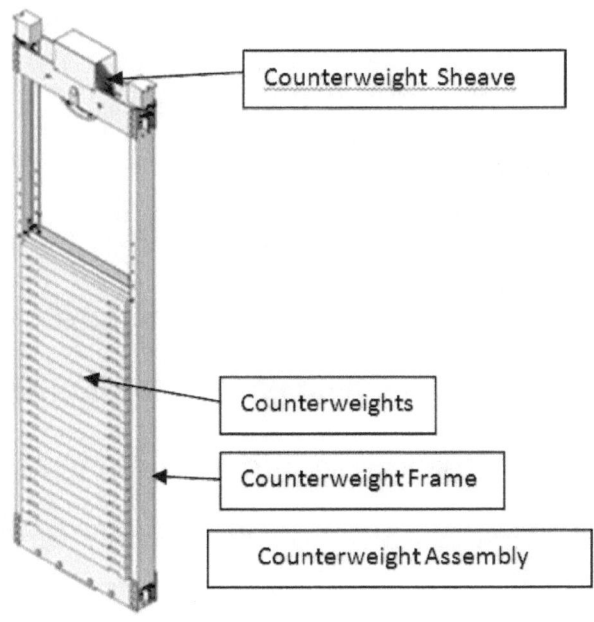

Counterweight Sheave

Counterweights

Counterweight Frame

Counterweight Assembly

Counter weight is rectangular in shape and moves up and down the same hoistway. Since the car and counter weights are tied to the opposite ends of the rope, the counter weight moves down when the car moves up and vice versa.

1.9 Why the car and counterweight tied by a rope on either side of the pulley do not fall?

One end of the rope is attached to the elevator car, looped around a sheave (pulley) and the other end tied to the counterweight frame. A sheave is just a pulley with groove around the circumference. The sheave grips the hoist ropes, so when you rotate the sheave, the ropes move too. The friction between the ropes and the sheave generates the traction which gives this type of elevator its name. The deep groove in the sheave provides the necessary traction and helps to hold the car and counterweight in their positions without freewheeling. This traction is the very basis of elevator technology.

1.10 What is the material used in a rope?

The car is raised and lowered by traction steel ropes. Recently few elevator manufacturers have replaced the ropes with Coated Steel belts. However we shall continue with the concept of roped elevators which are still very popular worldwide.

1.11 How the rectangular car moving inside the shaft does not tilt, spin etc.?

On either side of the car, guide shoes are fitted which guide through the guide rails which are fixed on the walls of the elevator shaft using brackets. The similar arrangement is made to the counterweight (CWT) frame also for smooth movement of CWT frame up and down.

Guide rails are necessary on either side of the car for following reasons:

To guide the car in vertical travel and to prevent horizontal movement or lurching of the car/ CWT as much as possible, and to prevent tilting of the car due to eccentric load, such as, when a group of passengers stand on one side or rear of the car.

When a car is loaded with passengers, there is an eccentric load on the car; it is prevented from tilting by the guide shoes (or rollers guides) pressing on the rails. The rail now acts as a beam supported by the brackets and it must have sufficient strength to carry these forces and also have sufficient stiffness to keep the front edge of the platform level with the landing as loads enter or leave the car. In the case of passenger elevators, where the eccentric loads are small these properties are not that important as with freight elevators where eccentric loads are usually very large.

Greater the load, larger is the required cross sectional area of the guide rail. In addition to the cross sectional area, it must be supported at certain regular intervals or it tends to buckle, if the points of support are too far apart. The distance between the points of support to prevent buckling depends on the moment of inertia and on the cross sectional area of the guide rail.

The guide rail has also got a very important function to stop and hold the car in the event of free fall of the car. The guide rail acts as a column during safety application.

1.12 How the car and Counterweight (CWT) move up and down?

The sheave which holds the car and the counterweight on either side, is connected to the shaft of an electric motor .When the motor turns one way, the sheave raises the elevator car, when the motor turns the other way, the sheave lowers the elevator. In the earlier designs of electric elevators, 3 phase induction motors and gears with or without speed controls were used. But in recent designs Permanent magnet DC motors are used.

1.13 What is the normal supply voltage to the motor?

The normal power supply is 3 phase, 400/415 Volts 50 Hz. We shall start our discussions with 3 Phase induction motors and later move on to PM AC motors and their speed controls. In India, 3 phase induction motors typically work on 400/415V, 50 Hz power supply. Elevator motors typically used to have 4 or 6 Poles.

1.14 How is the rated speed achieved?

We know that the RPM of the induction motor is determined using the formula 120 f / P, where f stands for frequency of power supply and P stands for the number of poles in the motor. Using the above formula, the rpm of the 4 pole and 6 pole inductions motors work out to 1500rpm and 1000 rpm,(25 and 16.66 revolutions per second respectively).

The minimum diameter of the ropes allowed as per Indian standards is 8 mm and the Indian standard specifies that the minimum diameter of the driving sheave has to be 40 times that of the rope to get the optimum wrap angle. By this code, the minimum diameter of the sheave works out to 320mm and a circumference of 1006mm, which means for one revolution of the sheave the lift will travel 1.006 meters. From our calculation in earlier Para, the 6 pole motor makes 16.66 revolutions per second, making the elevator to travel at a speed of 16.75 mps. But, even one of the fastest elevators in the world at Burj Khalifa runs only at 10 mps. Typically, a residential elevator runs with a minimum of 0,63mps and max of 2.5mps. The speed of the motor is reduced using gears.

1.15 What is the function of a Gear in elevators?

To achieve this speed, we require a reduction gear coupled between the motor shaft and the driving sheave.. The worm gear used in elevators, not only decreases the rotational speed of the traction pulley, but also change the plane of rotation. By decreasing the rotation speed, with the use of a gear reducer, we are also increasing the output torque, therefore, having the ability to lift larger objects for a given pulley diameter. A worm gear is chosen over other types of gearing possibilities because of its compactness and its ability to withstand higher shock loads. It is also easily attached to the motor shaft, sometimes through use of a coupling. The gear reduction ratios typically vary between 22:1 and 69:1.

Summary:

The elevators achieve their vertical motion from an electric motor. Steel cables or hoist ropes run from the top of the elevator car, over the drive sheave to the top of the counter weight. The downward force caused between the gravity acting on the weight of the car and counterweight creates friction between the steel cables and the driving sheave, thus creating traction. As the sheave rotates, the elevator car is raised or lowered. The car and Counterweight guide rails guide the movement of the car. Rails also provide safety in case of free fall due to breakage of ropes, by action of safety block mounted on the car and the rail.

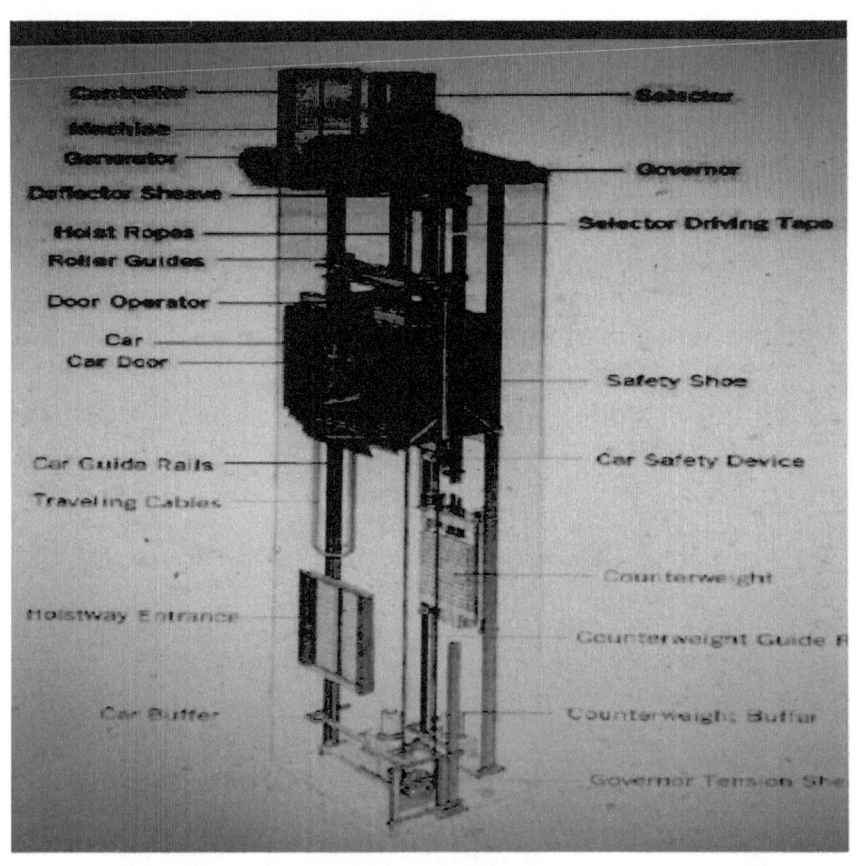

Courtesy: Project submitted by Loi Cheng
for Rensselaer Polytechnic Institute – May 2011

CHAPTER 2

EVOLUTION OF CONTROL AND DRIVE SYSTEMS
IN ELEVATOR DESIGN

2.1 Sequence of steps for the elevator to run:

We shall pick up the elevator control and drive designs which existed in India during the mid-1980s and proceed from thereafter.

A basic elevator control and drive system should have the following minimum intelligence:

- A means to know the position of the car.

- A means to upgrade the car's position while moving.

- Should be able to receive calls from the car inside or the landing.

- Close the car and landing doors and move

- Decide the direction of travel by comparing the position of lift car with called floor..
- Stop at the required floor
- Open the doors.

2.2 How does the elevator sense its position in the hoist way?

The car is mounted with many mechanical cams and they operate switches mounted in the hoist way. The control system looks for operation of SET switch which is mounted at the bottom most floor and sets its own position as Zero. If the control system does not find the signal, it means the car is above that floor. In that case the control system makes the car to move down at a very slow speed till the SET switch is operated. This has been the logic followed right from 1980s till now. Later designs had a battery backup to keep its position with power failure. But operations of SET switch at the bottom most floor was the most reliable way of knowing its position even today. In the recent designs, the lift car retains its position even if the power is turned OFF and ON thus avoiding correction run to the bottom floor.

2.3 How does the elevator upgrade its position during movement?

There are many methods followed. One popular method is to fix a metallic vane at each floor and a photo electric switch is mounted on the car top. As the car moves, interceptions of vane with the photoelectric switches are counted by the control logic to update its position. During the UP travel the counts are increased and while traveling down the counts are decreased.

Photo Electric Switch

Each time the car reaches the ground floor, the count is reset to Zero by virtue of the SET switch to take care of any possible counting errors.

2.4 How does the elevator know that a call has been registered?

Elevator control system is generally nstalled in the machine room. This controller is microprocessor operated. The call buttons in the COP (Car operating panel) mounted inside the car are wired to the controller. Similarly, there are hall buttons in each landing which are also wired to the controller. Whenever a Car or Hall button is pressed it is sensed by the controller. In the newer designs discrete wiring of these buttons has been replaced by serial communication. Serial communication saves cost due to reduction in the number of wires, especially as the number of floors increase.

2.5 How does the car move to the required floor?

The controller compares the position of the lift with

Hall Button Assembly

respect to its call. For example, if the lift is in position 3 and a COP button corresponding to floor 8 is pressed, the controller decides the direction of travel by comparing the position of lift with called floor: This is achieved by the control logic by comparing the car's position with that of the called floor. First the controller commands to Close the door. Before actual movement of car, the doors must be closed. Where manual doors are used, it has to be physically closed by the passenger. In case of auto doors, the controller signals the door operator, which is mounted on the car to close. The car door is closed by means of door operator control logic. The landing doors are mechanically coupled to the car door and hence the landing door closes along with the car door. Both the car and landing door must be closed before a run is initiated. This is sensed by operation of independent switches for car and landing doors. When the car door closes, the car and landing switches operate and the car can now move.

As the lift moves, the photo electric cells are intercepted and the controller increments its count. When the count matches with the called number of the call button, the controller makes the motor to slow down and stop at door zone. At each floor level, another photo electric switch mounted on car top is intercepted by a vane mounted in the hoistway known as door zone switch. Once the door zone is reached and if the car is moving at slow speed, a stop command is given by the logic control to the motion control and the car stops at the door zone. After stopping, the controller gives a command to the door operator to open the door.

2.6 Classification of Elevators:

For easy understanding, the author has classified the elevators based on their different attributes.

2.6.1 The Indian Standard classifies the type of elevators as Passenger lift, Goods lift, Hospital lift and Service lift. This classification is made for the purposes of outlining the speed and load.

2.6.2 Geared and Gearless Elevators :

2.6.3 Elevators with machine room and machine room less (MRL) elevators

2.6.4 Elevators based on type of drive system namely Single Speed, Two speed and variable speed.

2.6.5 Elevators based on the type of control system namely Down Collective, Simplex, Duplex, Triplex and multiple car elevators

2.6.1 **Classification as per Indian Standard**

The fundamental specification of an elevator is the load it can carry in Kg and its rated speed in meters per second m/s. The other equipment is built to fulfill these parameters and normal operation is then guaranteed by the manufacturer.

The Bureau of Indian Standards IS 14655 specifies the weight of one passenger for the purposes of any calculation as 68 Kg. The European Standard as per

EN 81 specifies the load of one passenger as 75 Kg for the purposes of calculations.

2.6.1.1 Passenger Lift: A lift designed for the transport of passengers.

2.6.1.2 Goods Lift: A lift primarily designed for the transport of goods but which may carry a. lift attendant or other person necessary for the unloading and loading of goods.

2.6.1.3 Service Lift (Dumb Waiter): A lift with a car which moves in guides in vertical direction, has net floor area of 1 sqm, total inside height of 1.25m whether or not provided with fixed or removable shelves and capacity not exceeding 250Kg and is exclusively used for carrying materials and shall not carry any passenger.

2.6.1.4 Hospital Lift: A lift normally installed in a hospital/ dispensary/ clinic and designed to accommodate one number bed /stretcher along its depth with sufficient space around to carry a minimum of three attendants in addition to the lift operator.

2.6.1.5 The table below outlines the recommended Load and Speeds for a maximum travel of 30m. The code also provides the area required inside the lift for the passengers to stand comfortably in a fully loaded car.

	Passenger Lift	Goods Lift	Hospital Lift	Dumb Waiter
Typical Speed (mps)	0.7 ~2.5	Up to 0.5	Up to 1.5	Up to 0.5
Load in Kg ('P' indicates No of	272 (4P) 408 (6P) 544 (8P) 680 (10P)	500 1000 1500 2000	1020 (15P) 1360 (20P)	100 150 200 250

Passengers)	884 (13P)	2500	1768	
	1088(16P)	3000	(26P)	
	1360(20P)	3500		
		4000		
		4500		
		5000		

Table 1: Recommended Load and Speeds as mentioned in IS14665 (Part 1): 2000

2.6.1.6 Recommended car inside area: (Refer BIS)

No. of Passenger	Rated Load @68kg/ Pass	Max Net Inside Car Area	Proposed Allowable Minimum Area
4	272	0.77	0.68
5	340	0.95	0.85
6	408	1.12	1
7	476	1.28	1.16
8	544	1.45	1.31
9	612	1.6	1.45
10	680	1.76	1.6
11	748	1.91	1.74
12	816	2.05	1.87
13	884	2.2	2.01
14	952	2.34	2.14
15	1020	2.47	2.27
16	1088	2.61	2.4
17	1156	2.74	2.52
18	1224	2.87	2.64

19	1292	3.00	2.76
20	1360	3.13	2.88
	1500	3.38	3.11
	2000	4.22	3.88
	2500	4.99	4.59

2.6.2 Geared and Gearless Elevators:

- In gearless elevators, the motor rotates the sheaves directly.

- In geared elevators, the motor turns a gear that rotates the sheave.

Let us first understand why the gears were required in an elevator. Elevators use Worm gears to reduce the speed of the motor to drive a large gear wheel to turn with more force (Enhanced torque). It is also useful for changing the direction of rotation in gear-driven machinery. As mentioned in chapter 1, a worm gear is chosen over other types of gearing possibilities because of its compactness and its ability to withstand higher shock loads. It is also easily attached to the motor shaft, sometimes through use of a coupling. The gear reduction ratios typically vary between 22:1 and 69:1.

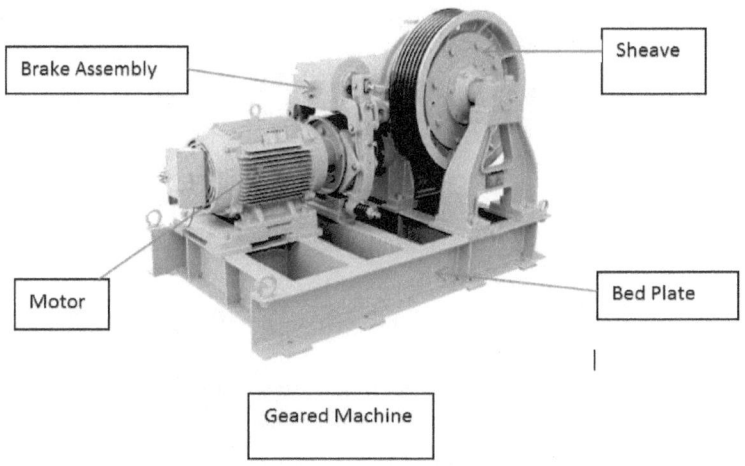

Geared Machine

By proper selection of gear ratio and the sheave diameter it is possible to attain speeds up to 2.5 mps from a 1000 or 1500 RPM induction motor.. For speeds above 2 mps, it is preferable to use Gearless lifts. The major components of a geared machine are motor, brake, worm gear, Traction sheave and brake assembly. The control system,

Governor, Interphone, Automatic Rescue systems etc are all housed in a **machine room** above the elevator shaft.

The motors used are normally 3 phase induction motor. When compared to other applications, in lifts, the motor is to start with load. Normal motors have 150% to 175% starting torque. Since the lift motors have to start with the load, anything between 250% to 300% torque is required. Motors for elevator applications are made with this specific requirement. With advancement in power electronic technologies, the high starting torque requirements are controlled by the electronic drive system, the ride quality and leveling accuracy have improved to a large extent along with the efficiency.

In order to obtain the benefits of gearless system, it was required to reduce the speed of the driving motor rpm to very low levels which was technologically difficult till recent times. The earlier types of gearless elevators used Ward Leonard speed control system, wherein a Generator was driven by a prime mover and the generated power was used to control the speed of the DC motor by controlling its armature current.

The technological advances have made it possible to control the speed of the Permanent magnet synchronous motors by Pulse width modulation techniques. Permanent magnet motors are designed to have very low speed and with high torque. The machines come with wheel and brake all assembled as a package.

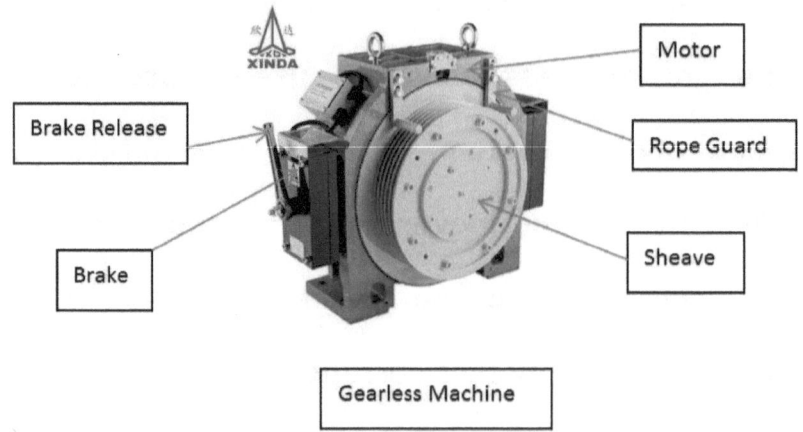

Brake Release

Motor

Rope Guard

Brake

Sheave

Gearless Machine

The geared system has the following disadvantages over the gearless system

- Less Efficient due to losses

- More power consumption due to energy loss

- Require oil for cooling the gears

- Requires maintenance.

- Additional Space

- Poor ride quality such as noise, vibration.

The author considers a gearless traction machine superior to a Geared machine because it is nearly 40% more efficient and quieter in operation, requires less maintenance and has longer life, no bearing noise, less vibration, no oil for cooling the gears and hence no problem of oil leakage, . The decision as to whether these advantages are worth the additional cost involved can be made only after careful analysis. Generally a gearless machine was preferred for high rise, high speed operation but with the advent of technology the gearless elevators are now used for even up to 1 mps. For Speeds below 1mps, the geared systems are still popular, though fading.

2.6.3 Elevators with machine room and machine-roomless elevators:

At present the elevator industry is offering a relatively new elevator product termed Machine Room-Less Elevators (MRL's) application in new construction or major renovation projects compared to standard elevator equipment. It is a fairly significant decision as it will affect the design of the elevator hoistways and equipment rooms.

The machine-room-less elevator is the result of technological advancements that often allow a significant reduction in the size of the electric motors used with traction equipment. These newly designed permanent magnet motors (PMM) allow the manufacturers to locate the machines in the hoistway overhead, thus eliminating the need for a machine room over the hoistway. This design is nowadays becoming the standard product for low to mid rise buildings. It was first introduced to Europe market by KONE though with geared arrangement. Product acceptance was initially slow in the U.S. market because of its initial, limited applications, its inability to meet U.S. code requirements, and the limited number of manufacturers offering an equivalent product.

In the past few years the elevator manufacturers have overcome the obstacles to acceptance of the MRL product. All of the other elevator manufacturers are now marketing their versions of the MRL, and the product offering has been expanded to include most of the more popular elevator sizes and speeds. In addition, local code officials have become more receptive to the change in technology.

Recent developments by introduction of coated steel belts instead of ropes have made it possible to reduce the sheave diameter substantially to get 1mps speed at about 490rpm with 2:1 roping. The coated steel belts are used only by a few manufacturers and the steel ropes are still popular with most of the manufacturers.

While the hoisting motor is installed on the hoistway side wall, the main controller is installed on the top floor next to the landing doors. This controller is situated behind a locked cabinet which has to be unlocked using a key for maintenance, repair or emergency purposes. Inspection windows were provided to have visibility inside the shaft especially the main rope the governor and Governor ropes. Most elevators have their controller installed on the top floor but fewer elevators have their controller installed on the bottom-most floor. Some elevators (like OTIS and Schindler) have the controller cabinet installed within the door frame instead on the wall to save space. In some cases, Otis motors are installed above the counter weight rails. There is a continuous improvement happening in the positioning of the various equipment considering the maintainability and safety

Like normal traction elevators, M.R.L. elevators use the conventional steel cord ropes used as the hoisting cables. Some elevator brands (such as Otis, Schindler and ThyssenKrupp) are using flat steel rope belts instead of conventional ropes. Manufacturers using this technology claimed that with flat steel belt ropes, it saves much space on the hoist way and to allow a minimum size of the hoisting sheave.

Most M.R.L. elevators are used for low to mid rise buildings. M.R.L. elevators in mid-rise buildings typically serve up to 20 floors.

2.6.3.1 Benefits of the MRL Elevator:

- Machine room is no longer needed, as all machineries successfully fit into the hoist way, except the control panel, which can be placed anywhere within a short distance from the traction machine.

- A conventional machine room elevator applies load stress on a building's structure, whereas

47

the guide rails support as much as 75% of the stress for added building friendliness in case of certain Otis and Mitsubishi elevator types.

- In new building applications, Architects naturally like to adopt the fact that no rooftop lift machine room is required allowing a clean roofline and maximizing available space

2.6.3.2 Limitations of the MRL Elevator:

- As previously stated, the machine room less lift depends on the relocation of machine room parts in order to negate the need for a separate room. This means that a new position for the control panels needs to be found. In most cases the panels are incorporated into extended architraves or special panels, usually at the top floor. Not only must the visual aspects of the panels be accepted by the Client, their position increases the maintenance necessity at this landing. In an office block this could cause some disturbance, but if children or animals are resident, it could cause an increased risk to the safety of all, including the maintenance engineer.

- In certain cases, especially with five star hotels, the author has attended complaints of elevator noise in the adjacent rooms due to machine noise and / or noise from the controller parts.

- As far as the user is concerned, there is very little difference under normal conditions between a machine room less lift and any other traction drive unit. However because the winding machine is located inside the shaft it is not directly accessible.

- During a power failure, in case of elevator with machine room, it can be manually cranked up or down so that passengers can be released from the nearest floor. Under normal circumstances the car would be moved in the direction of least resistance determined by whether the car or counterweight is heavier. Should circumstances dictate that the car is moved in a particular direction, for example a fire on the floor below a fully loaded car, this can be achieved, if perhaps, it takes a little effort. Where the drive machine is not directly accessible, the full control in an emergency is not possible. On some machine room less applications a battery backup system can be provided, as an extra cost, but it often cannot move a fully loaded car

- When the elevator stops suddenly due to power outage, the MRL Gearless elevators seem to introduce heavy jerk as compared to equivalent geared elevators. This causes discomfort especially to elderly and pregnant women.

- Until the very recent past the MRL elevator products did not offer an economical advantage over traditional elevator products. The market is now experiencing certain MRL elevators priced more competitively, comparable to a traditional geared traction elevator. Certain manufacturers have tried to introduce a cost competitive product to compete with the low rise hydraulic elevator.

2.6.4 Elevators based on type of drive system namely Single Speed, Two speed and variable speed

2.6.4.1 Single Speed Elevators:

These are the most common type of low-capacity, slow-speed, low cost passenger elevators intended to service buildings in the two to eight floor ranges. These are elevators at 0.63 mps where car speeds and passenger comfort are less of an issue. In these types of elevators motor power is switched directly on line and stopped by the removal of electric power and application of brake simultaneously. In certain types, a step of resistance may be introduced between the motor and the lines to reduce the inrush current when the motor is started and the resistances shorted when the motor attains normal speed. With this arrangement, the starting jerk is reduced to a certain extent. These types of elevators are still most popular in low cost, low rise residential buildings.

For lifts up to 0.63 mps and 550Kg capacity, the Indian standard allows leveling inaccuracy of +/- 75mm (max).

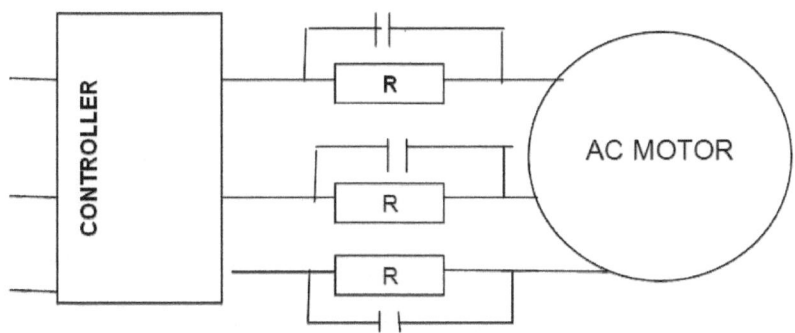

Disadvantages:

1. Poor ride Quality – Acceleration, Deceleration, Vibration and stopping Jerk.

2. Poor leveling accuracy - In this type of elevator control, it is extremely difficult to consistently bring the car to a stop in the same spot where it is level with the floor. There are several

reasons for this inconsistency. Varying load conditions affect the ability of the brake to stop consistently at level. The brake tension springs have been set for a "typical" loading condition and then set to stop the car level with the floor. A heavier load will cause the car to slide a little further (travel past the floor) when it stops and a lighter load will cause the car to stop a little sooner (stop prior to floor level). The counterweight is always about 40% to 45% heavier than the empty car, thus an empty car will have a different stopping condition than a fully loaded car.

3. The speed of the elevator gets affected by the slight variations in the line supply frequency or voltage, thus affecting the leveling accuracy.

4. Tripping hazard due to poor leveling accuracy.

5. Not suitable for wheel chair applications.

6. There is no motion control circuitry for re-leveling.

2.6.4.2 Two Speed Elevators:

Further refinements of alternating current motor control include the use of two speed motors with ratios of 4:1 between the low speed and high speed windings. The speed of the elevator could be increased from 0.63mps to 1 mps with two speed elevators. The elevator runs with the high speed winding and is switched to the lower speed winding when it is near the floor stop. Final stopping is done with the brake, simultaneously removing the power to the motor. In order to reduce the starting currents and jerk during switching from fast to slow speed windings, resistors were introduced between the line and the respective motor windings which get bypassed once the rated motor speed is achieved. For lifts up to 1 mps and drives with two speed motors having a ratio of 1:4 between high speed and low speed are allowed max +/- 25mm leveling accuracy as per IS 14665 (Part 3 / Sec 1):2000 para 10.

Disadvantages:

Two speed elevators have all the disadvantages of Single speed elevators. **These types of elevators have become obsolete in India** owing to the availability of cost effective better speed control methods.

2.6.4.3   Variable Speed Control:

2.6.4.3.1 Ward – Leonard Speed Control:

The Ward Leonard control system was widely used for elevators until thyristor became available in the 1980s, because it offered smooth speed control and consistent torque. The speed control of DG motor accomplished by means of an adjustable

voltage generator is called the Ward-Leonard
system.

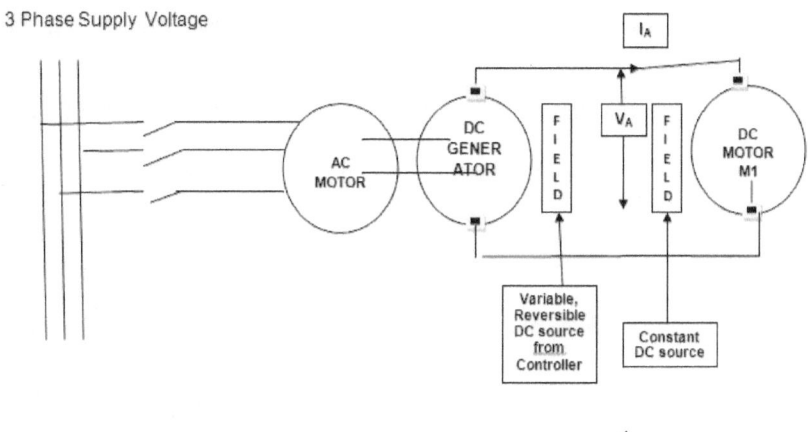

The Ward Leonard configuration consists of a
constant flux DC motor and a rotating conversion AC
motor DC generator set. As shown in fig above, M1 is
the main motor whose speed control is required. The
field winding of this motor is permanently connected
to DC supply and armature is fed from variable
voltage so that the motor can run at any desired
speed. To provide this variable voltage, a motor DC
generator set is used in which the generator is

directly coupled to a constant speed motor. The armatures of the generator and the motor are directly electrically connected. The field of the generator is excited from a variable and reversible DC source thus the generator output voltage can be varied from 0 to maximum in both directions. This system of speed control is commonly used in elevators as this can give speed control in either direction. Since the generator output voltage can be increased from zero, no extra starting equipment is required to start up the main motor smoothly. Ward Leonard system is a classic high quality elevator drive arrangement found in the vast majority of better quality geared and gearless installations built before 1990.The important feature of the Ward Leonard system is its regenerative action.

The fundamental principles of DC variable speed drive, with shunt wound DC motor are relatively easy to understand covered by a few equations as follows:

The Armature Voltage 'V_A' is equal to the sum of Back emf of Motor 'V_E'and the voltage drop due to armature current flowing through the armature resistance.

- Hence Armature Voltage $V_A = V_E + I_A * R_A$

The DC motor speed 'n' is directly proportional to the motor back emf 'V_E' and inversely proportional; to the field flux '$\emptyset$'. The flux is proportional to the field current 'I_E'.

- Motor Speed n $\propto V_E / \emptyset$ Motor speed can be varied either by varying V_E or $\emptyset$

- The output torque of the DC motor ' T ' $\propto I_A * \emptyset$

By changing the direction of Armature current or by changing the direction of filed current, the direction of rotation or the direction of the torque can be changed.

- The output power of the motor $P = T * n$

From the above, following conclusions can be derived:

➢ The speed of the DC motor can be controlled by adjusting either the armature voltage or the field flux or both. For elevator applications the field flux is kept constant and the motor speed is varied by increasing the armature voltage.

➢ When the Armature voltage V_A has reached the maximum voltage, it is possible to increase the motor speed by reducing the field flux. This is known as field weakening method.

➢ Since the torque is not dependent on V_A , the full load torque output is possible over the normal speed range even at stand still (Zero Speed)

➤ The output power is zero at zero speed. In the normal speed range at constant torque, the output power increases in proportion to speed.

➤ In the field weakening range, the motor torque falls in proportion to the speed. Consequently the output power of the DC motor remains constant.

Torque and power characteristics of a DC motor over its speed range is shown below;

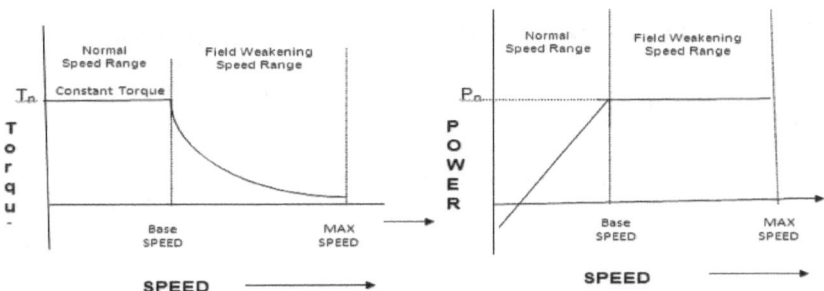

Advantages of Ward Leonard system:

- A wide range of speed from stand-still to high speed in either direction.

- Has good speed and torque characteristics and could achieve a speed range of 25:1

- Extremely good speed regulation.

- Regenerative action

Disadvantages

- High initial cost due to motor generator set

- Occupies more space

- Low overall Efficiency.

- High maintenance cost primarily due to wear and tear of commutator and brushes.

- Mechanical commutator and brushes impose restrictions on ambient temperature and humidity.

- High thermal losses in the machine room

- High Noise levels

Similar to 2 speed elevators, these types of elevators also have become obsolete in India owing to the availability of cost effective better speed control methods.

2.6.4.3.2 ACVV (AC Variable Voltage) Speed Control:

In this method, the voltage applied to the three phase windings of low cost squirrel cage induction motors were controlled with the help of thyristors. The deceleration was achieved by DC injection into the windings. The resultant speed control is smooth and step less, making it suitable for passenger elevators.

Demerits:

- High motor slip caused by the constant frequency of the variable voltage causes large thermal losses with resultant low operating efficiency.

- The chopper which is used to provide the variable voltage control introduces undesirable harmonics into the power system thus causing malfunctions to the other electronic systems connected to the same power system in the building.

- These harmonics cause undesirable radio noise and can cause system component overheating.

- The system's low power factor increases line losses and necessitates increased feeder sizes.

As a result of the demerits, this speed control system which was applicable to low and midrise passenger service with speeds up to 1.75mps has been replaced by VVVF controls. The ACVV elevator speed control has now become obsolete.

2.6.4.3.3 ACVF / VVVF Speed Control:

Both the voltage and frequency of the motor windings are varied to control the speed of the motor. IGBT (Insulated Gate Bipolar Transistors) are used to accomplish the voltage and frequency changes. Since the speed of the induction motor is proportional to the frequency of the supplied power, the speed of the motor is varied. Both the voltage and frequency are varied to control the torque and current.

The system consists of a rectifier, which changes the incoming AC to DC, and an inverter which creates variable voltage and variable frequency three phase AC from the rectified DC voltage. This output is then applied to a standard squirrel cage induction motor which operates essentially at the speed corresponding to the frequency. By maintaining a constant frequency to voltage ratio input to the traction motor, it is possible to provide continuously variable highly accurate speed control throughout the full speed range of the motor. The VVVF speed control system eliminates most of the disadvantages of the previous ACVV thyristor control system and has the following characteristics:

- Overall system efficiency high at all motor speeds

- Use of Economical commercially available Squirrel cage induction motors.

- System power factor closer to Unity.

- Line harmonics are much less than ACVV controls.

- Extremely good speed control and leveling accuracy.

- Uses solid state, thus very low maintenance

- Electric energy use and peak loads are reduced thus reducing energy bills.

- Brake applied on reaching near zero speed, thus reducing jerk and brake wear and tear.

- The system is suitable for all rises and speeds.

- Machine rooms are smaller, cooler and quieter.

2.6.5 Elevators based on the type of control system namely SAPB, Down Collective, Simplex, Duplex, Triplex and multiple car elevators

2.6.5.1 SAPB (Single Automatic Push Button Control):

This forms the simplest type of control system. In this system, all floors have only one hall button. The buttons may or may not have telltale lights(TTL) to indicate call registration. At a time only one call will be attended. The control system does not register calls while running. It attends to the first person who presses the hall button after completing a travel. This type of arrangement is used only in single speed elevators with low cost applications.

2.6.5.2 Down Collective control:

For this system, all floor have DOWN hall buttons only except the ground floor where there is UP hall button only. If there are basements, Ground floor will have both UP and DN hall buttons and the basements have UP hall button. The operation is similar to Simplex and Duplex controls. Since there are only down hall buttons all the calls are serviced sequentially during the down travel of the elevator. These types are most suitable for residences and small hotels where inter floor traffic is very rare. This type of arrangement not only reduces cost but also reduces unnecessary waiting time in residential elevators.

2.6.5.3 Simplex Full Collective Control:

This is a fully automatic operation used for a single elevator system. Hall fixtures have two buttons, one for UP travel and the other for DOWN travel. If our intended direction of travel is in up direction, UP button has to be pressed and if our intended direction of travel is in down direction, the DOWN button has to be pressed. In acknowledgement to our pressing of the button, a light inside the button glows and this is called a telltale light. Hall calls in the direction in which the elevator is traveling are responded sequentially and when calls in that direction are cleared, calls in the opposite direction will be responded to. When there are no more calls, the elevator will stay on the last attended floor level. In some cases, the free car is moved to ground floor where the arrival of passengers is more frequent. With the advent of microprocessor technology, parking of free car at any predetermined desired floor is possible.

2.6.5.4   Duplex Collective Control:

This is a fully automatic operation used for two elevator systems. Hall fixtures have two buttons, one for UP travel and the other for DOWN travel. If there are buttons for each elevator, as in simplex control, it is called **two Riser** arrangements. Most common type of arrangement of hall buttons is of **one riser**, meaning, one set of UP and DN buttons at each floor common to both the elevators. This type of arrangement saves cost. The operation is similar to simplex except that the calls are shared by both the elevators and hall calls are responded to by whichever elevator that can serve the hall call faster. In this type of arrangement, both the elevators share a common machine room and the elevators are located adjacent to each other.

2.6.5.5   Group Control:

This is a group control system used to operate a bank of three or more elevators, normally limited to six though it is not uncommon to see eight elevators in a group. The elevators share a common machine room and are located near each other. The numbers

of risers have to be decided based on the location of

elevators and convenience of the users.

2.7 Typical Arrangement of Control System:

2.7.1 Description of Control system in an Elevator:

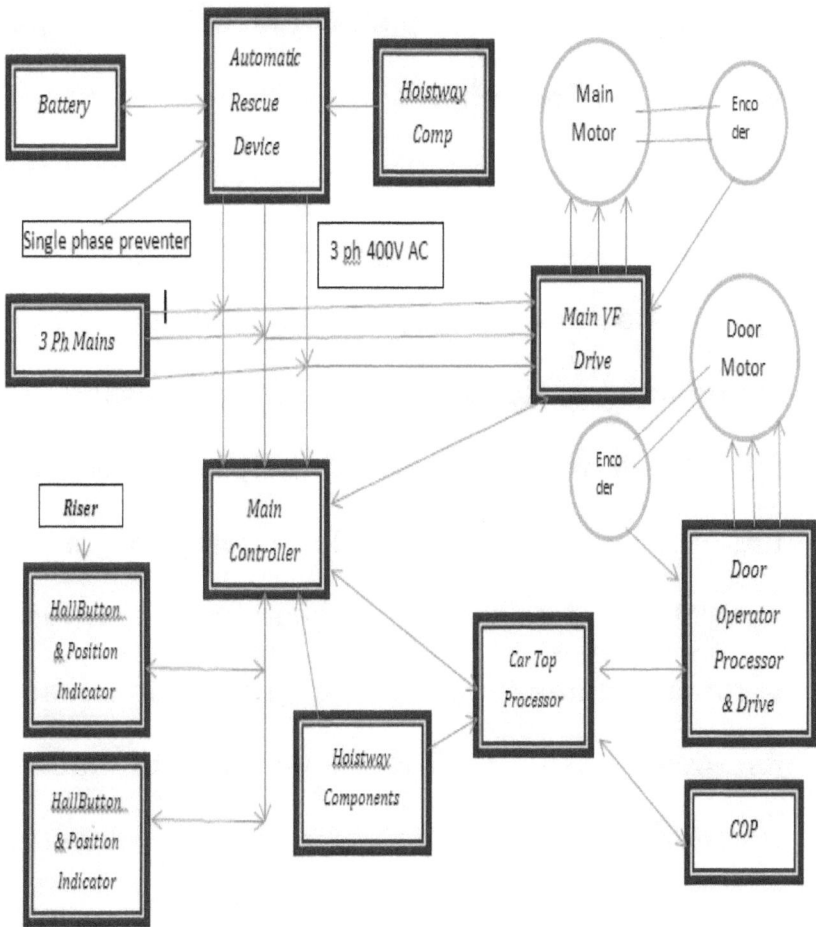

In actual practice, the arrangement of electronic assemblies and their inter communication vary from manufacturer to manufacturer. However a general popular arrangement of electronic assemblies and their inter communication has been described in the above picture and below paragraph.

2.7.1.1 Main Controller:

The main controller is the heart of the elevator system. Initial controllers were designed using electro mechanical relays but most of the controllers manufactured today use micro-processors. The controllers have two parts namely logic control and the motion control. Basic operation of logic controllers is to receive signals from car or hall buttons and decide the direction of motion based on its own position. The command to move is transferred to the motion control which in the current designs is the ACVF drives. ACVF drives provide a smooth start, control the speed of the main machine and stops the elevator without jerk. The ACVF elevator drives are smart devices to control the motion and provide a smooth ride to the passengers inside the elevator. Smooth speed control is achieved due to the closed loop feedback provided by the encoder mounted on motor shaft. Switches mounted in the hoistway provide the door zone, limit and floor position signals. As the car moves, the floor position signals are counted by the logic controller to update its position.

2.7.1.2 Hall Button & Position Indicator:

This panel which is mounted in each floor, near the hoistway door, has an UP button, a DN (Down) button and an optional position indicator. The UP button must be pressed, if one wants to go up. Similarly the Down button must be pressed, if one wants to travel down.

In general all the Hall buttons are connected serially to the Processor board in the machine room. Whenever a button is pressed, it is communicated to the master processor in the controller. It is also common to note that in low rise buildings, the hall buttons are discretely wired to the master control unit. Communicating serially or in parallel is a commercial decision, however nowadays most of the elevators operate on serial communication. In general, the hall fixtures communicate with the master controller either in RS485 , CAN or RS422 communication protocol.

Riser: Each floor has a Hall Button and Position Indicator. This vertical arrangement of HB assembly is called a Riser.

2.7.1.3 Hall Lanterns & Gong:

Hall lanterns and gong are optional fixtures opted in multiple car groups to announce the arrival of an elevator to a particular landing and its subsequent direction of travel. These fixtures also communicate with the master controller and get the commands.

Hall Lantern & Gong

2.7.1.4 Car Operating Panel (COP):

The calls can get registered from the car operating panel also. Whenever a call is registered, it is immediately acknowledged by the controller by lighting the call button, which is termed as tell-tale light. The tell-tale light feature is available in hall buttons also.

2.7.1.5 Movement of Elevator

On entering the lobby of any building, one finds elevators; single, two or multiple cars in the lobby. There are fixtures with UP and DN buttons. The UP button is to be pressed if one intends to travel in up direction. Similarly, the DN button is required to be pressed if the intended travel direction is DOWN. Apart from the Hall button fixture, one may also notice visual indicating displays showing the actual location of the lift at that time and the direction of travel. The controller decides the direction of ravel and commands the ACVF drive accordingly. The ACVF drive ensures the speed and ride quality such as acceleration and jerks.

In some lift lobbies, the hall fixture may have only one button. The control system in this case is called Down Collective Control (DCL). The elevator stops to the called hall button, only during its down ward travel.

CHAPTER 3

BASICCONSIDERATIONS FOR SELECTION AND SPECIFYING OF ELEVATORS

3.1 Fundamental Considerations:

One of the many decisions that must be made by the designer of a multi storey building is the selection of vertical transportation equipment that is passenger, service and freight elevators (and escalators). The quality of elevator service is an important factor in the tenants' choice of space in competing buildings. Although the final decision as to the type of equipment rests with the architect, the factors affecting it are so numerous that the building designer should consult with an elevator expert. This service available from the consultants and the major manufacturers of elevators should be availed. One of the purposes of this book is to familiarize the architect and building engineers to make preliminary design decisions and interact effectively with consultants / manufacturers.

When considering traffic design of a new building the major building dimensions must be known. Unfortunately it is often the case that the architect responsible for the building conception will have fixed the building core limiting the space available for the lift system or even will have defined the number of shafts, their dimensions and travel. This removes one very important degree of freedom from the lift traffic designer. The total flow of traffic coming into the building, vertical movement and those going out of the elevator decides the core efficiency of any building and hence it is essential that the architect consults the lift designer during conceptual finalization of any building.

Ideal performance of an elevator installation will provide minimum waiting time for a car at any floor, comfortable acceleration and deceleration; jerk free stopping, rapid transportation, smooth vibration free ride and accurate leveling at landings irrespective of load conditions in the car. The elevators must also provide smooth, quick, quiet and safe door operating system. Comfortable lighting, easy operation of car and landing buttons, floor level indicators to know our position inside the building, annunciators to announce arrival of cars, safe operation of all mechanical equipment under all load conditions & reliable security equipment, provide pleasant elevator use and experience.

Above all, the passenger safety is utmost important even under failure of any single component in any part of the whole elevator system and the failure is detectable by the service mechanic so that it gets fixed before another failure. Combination of two undetected failures may result in major safety issue. Elevator companies with strong design background ensure safety of the passengers by conducting design reviews and failure mode and effect analysis (FMEA) during design of equipment.

In addition to passenger satisfaction performance and service conditions, elevators also impact the architectural beauty of the building. The finish of cars and hoistway door and other equipment have to gel with the architectural beauty of the building. Integration of elevator hoistway spaces into the building and design of elevator lobby are also utmost important while selecting the finishes required in the elevator exteriors.

The elevators are covered by strict manufacturing and installation codes.

Many states in India have their own elevator codes guided by the bureau of Indian standards for elevators. There are codes related to fire safety, emergency evacuation, emergency power, handicapped elevators etc. Like most large industries, the elevator industry is self-regulating and standardized. Elevator company representatives are normally knowledgeable about all the codes and standards in force but this may not relieve the architect engineer of legal responsibility for the design. Therefore, it is strongly recommended that in the preliminary planning stage all pertinent regulations concerning vertical transportation be acquired, reviewed and understood.

Before any thought is given to the elevators in a building, a thorough and detailed study must be made of how people will arrive at the building, the building's occupancy and movement inside the building. Basic factors for determining the elevators for a building, the number of occupants in a building, the floor wise distribution of population, the arrival and departure rates at the lobby, the door performance namely the opening, closing and waiting time etc are required to be estimated. Also determination of elevators depend upon the type of building namely residential, office, commercial, Hotels, Schools etc. The actual use of the building is beyond the control of the architect or an elevator engineer as most of the factors are based on assumptions. Past experience and surveys of similar existing buildings can help confirm these estimates. **It is always advisable to make the conservative estimates as it would be difficult and very expensive to make changes in the elevator size, speed or number of elevators.** In many cases it would not be possible to alter anything due to lack of space. The entire building will lose its charm if there are long waiting lines near the elevators and if it takes considerable time for the residents to move from one floor to the other.

The National building code of India prescribes provision of elevators for buildings 15m or more in height.

3.2 Some General Considerations:

3.2.1 Planning arrival of passengers in lobby:

Initially the pedestrian movement planning considering the location of lifts, the entry and exit points, size of the lobby for a given number of elevators of specific capacity, Space for queuing have to be planned. Elevators must be placed, arranged and designed to provide the most cost effective performance. Unsatisfactory pedestrian movement planning can damage a building's reputation and cause loss in productivity for its occupants.

3.2.2 Lighting:

The interior of an elevator car should be well lit and the lights arranged so that they cannot be turned off by unauthorized persons. The EN-81 (Aug 2014) standard requires higher lighting levels in both the lift car and shaft. In car lighting should provide illumination of 100 lux (was 50 lux) with emergency illumination of 5 lux for one hour (was 1W for one hour). Lighting should be placed at a height of 1m in the centre of the car close to emergency push buttons. Inaccuracy in floor level accuracy can cause possible tripping hazard which can aggravate due to poor lighting.

3.2.3 Grouping of Elevators :

As a general rule, if elevator service is essential to the building operation, **two elevators** should be considered as minimum equipment in vertical transportation. With a single car installation, a passenger who misses the elevator has to wait till it makes a complete round trip. Any elevator requires periodic servicing and replacement of wearing parts. During that time, with a single car, one would go without service. Such a situation can cause utmost discomfort to all including sick and elderly. If a building requires more than one passenger elevator, it is recommended to group the elevators with two or more elevators in a group, owing to technological advances of the operating system, the wait time could be substantially reduced and one can board the earliest arrived car. The maximum number of elevators in a group should not exceed eight elevators due to complexity of the operating logic and non-availability of good operating system.

3.2.3.1 Two car Groupings:

For a two car group, side by side arrangement is the best. Passengers face both the cars and react immediately to an arriving car. Two cars facing each other is an acceptable alternative. In case of cars adjacent to each other, one can have single riser configuration, meaning common hall buttons, thus savings in one set of hall buttons. Also it would help having a common machine room so that inter connections between the elevator controllers become easy. Locating the elevators separately should be avoided. Separation of elevators at two different locations tends to destroy the group operation. The type of door may influence the choice. With center opening doors, adjacent cars may be preferable as it gives a balanced appearance. With two speed or single slide doors, unequal space between the two elevator entrances may make the elevator opposite to each other configuration more preferable.

3.2.3.2 Three car Groupings:

The arrangement of three cars in a row is preferable or two cars and one opposite is acceptable. Again, type of doors may influence the choice of location as explained for two car group.

3.2.3.3 Four car groupings:

Four elevator groupings are common in large busy buildings. Two opposite two is the most efficient option. All cars in a row arrangement increases the distance between the hall button and the last car in the row. With larger cars, the distance is appreciable considering the waiting population at the lobby.

3.2.3.4 Six car groupings:

Six cars, three opposite three is the most preferred arrangement. The waiting passenger can see all the elevators from his place and also can access all the cars easily. The lobby waiting time for elevators should be appropriately set to allow passengers to reach the elevators. Groups of six car elevators are very often found in large multi storied buildings.

3.2.3.5 Eight car groupings:

This is the largest group of elevators found in large buildings with four opposite to four being the most popular and practical arrangement. The main lobby must have opening at both ends so that the leaving passengers do not walk through the population due to other arriving cars.

3.2.3.6 Serving Floors:

All elevators in a group should serve the same floors. If for example, only one car out of a three car group serves the basement, people wishing to go to the basement from an upper floor have only one chance out of three. In such cases, it would serve better, if an additional elevator is used between the basements and the lobby floor.

3.3 Arrangement of Lifts:

The lifts should be easily accessible from all entrances to the building. For maximum efficiency, they should be grouped near the center of the building. It is preferable not to have all the lifts out in straight line and, if possible, not more than three lifts should be arranged in this manner. It has to be kept in mind that the corridor should be wide enough to allow sufficient space for the waiting passengers as well as for through passengers. The ideal arrangement of the lifts depends upon the particular layout of the respective building and should be determined in every individual case.

3.3.1 Passenger Lifts:

3.3.1.1 Low and medium class flats:

Where a lift is arranged to serve two, three or four flats per floor, the lift may be placed adjoining a staircase, with the lift entrances serving direct on to the landings. Where the lift is to serve a considerable number of flats having access to balconies or corridors, it may he conveniently placed in a well-ventilated tower adjoining the building.

3.3.1.2 Office buildings, hotels and high class flats:

It is advisable to have at least two lifts side by side at the main entrance and one lift each at different sections of the building for intercommunication. When two lifts are installed side by side, the machine room shall be suitably planned with sufficient space for housing the machine equipment. The positioning of lifts side by side gives the following advantages:

a) all machines and switch gear may be housed in one machine room,

b) the lifts can be inter-connected more conveniently from an installation point of view, and

c) Greater convenience in service owing to the landing openings and each floor being adjacent.

3.3.1.3 Shops and departmental stores:

Lifts in shops and stores should be situated so as to secure convenient and easy access at each floor.

3.3.1.4 For buildings with more than 12 floors:

it is recommended to have provision of 1 stretcher / service lift in addition to the passenger lifts. Where passenger and service lifts are provided in one lobby it is recommended to have group control for all the lifts.

3.4 Goods Lifts:

The location of lifts in factories, warehouses and similar buildings should be planned to suit the progressive movement of goods throughout the building, having regard to the nature of position of the loading platforms, railway sidings, etc. The placing of a lift in a fume or dust laden atmosphere or where it may be exposed to extreme temperatures, should be avoided wherever possible. Where it is impossible to avoid installing a lift in an adverse atmosphere, the electrical equipment should be of suitable design and construction to meet the environmental conditions involved. Normally goods lifts have lower speeds than passenger lifts for the same travel because traffic conditions are less demanding, and more time is required for loading and unloading. As loads for goods lifts increase in size and weight, so the operation of loading and unloading becomes more difficult. Therefore, it is usual to require greater accuracy of leveling as the capacity of the goods lift increases.

A large capacity goods lift at high speed is often a very uneconomical preposition. The inherent high cost is enhanced due to the very small demand for such equipment, much of which is custom made. The high capital cost of the lift, building work and electrical supply equipment usually shows a much smaller return as an investment than more normal sizes of lifts.

3.5 Hospital Bed Lifts:

Hospital bed lifts should be situated conveniently near the ward and operating theatre entrances. There shall be sufficient space near the landing door for easy movement of stretcher. It is convenient to place the passenger lifts in a hospital, near the staircases.

CHAPTER 4

PRELIMINARY DESIGN AND
TRAFFIC ANALYSIS

Contents mentioned in this chapter are furnished in detail in the Indian Standards document. Readers are advised to read the IS document.

4.1 Number of Lifts and Capacity:

The number of passenger lifts, their capacities and speed, required for a given building depend on the characteristics of the building. User satisfaction can be obtained only by providing sufficient number of lifts with adequate capacity and speed in order to reduce passenger waiting times.

Few important aspects to be considered are :

a) the number of floors to be served by the lift

b) the floor to floor distances to be travelled.

c) the population of each floor to be served, and

d) the maximum peak demand. This demand may be unidirectional, as in up and down periods, or a two way traffic movement.

In view of many variables, no simple formula is possible for determining the most suitable combination of lifts. It should be appreciated that all calculations on the traffic handling capabilities of lifts are dependent on a number of factors which vary according to the design of the building and the assumptions made on passenger movement..

In addition to the type of building, different authorities and manufacturers differ widely in their methods of calculation, due to variations in lift rates of acceleration and deceleration and door performance time. Therefore, the calculations made by different organizations will not necessarily agree.

It follows therefore, that the result of such calculations can only be put to limited use of a comparative nature.

4.2 Preliminary Lift Planning:

4.2.1 General:

Methods of calculating the traffic handling capabilities of lifts were first devised for office buildings. In due course detailed modifications were devised to suit other applications without altering the basic principles. The application to office buildings is still the most frequently used. Therefore, the following general method may be used as general guidance on preliminary lift planning for offices. A lift installation for office building is normally designed to service the building at a given rate and three main factors to be considered are;

a) population or the number of people who require lift service,

b) handling capacity or the maximum flow rate required by these people,

c) Interval or the quality of service required.

4.2.2 Population:

The first point to be ascertained from the eventual occupier is the total building population and whether occupier number is likely to increase in the future. If a definite population figure is unobtainable an assessment should be made from net area, and probable population density. Average population density can vary from about one person per 4 sqm to one person per 20 sqm. It is essential, therefore, that some indication of the probable population density should be obtained from building owner, If no indication is possible, 5 sqm per person for general office building can be generally assumed.

4.2.3 Quantity of Service :

The quantity of service is a measure of the passenger handling capacity of a vertical transportation system. It is measured in terms of the total number of passengers handled during each five-minute peak period of the day. A five-minute base period is used as this is the most practical time over which the traffic can be averaged. The passenger handling capacity should be approximately 10 percent to 15 percent of the estimated population that has to be handled in the building in five minutes for diversified tenancy office building and 15 percent to 25 percent for single purpose occupancy office building. For residential buildings, 7.5 percent is sufficient.

4.2.3.1 The handling capacity (HC):

The handling capacity of a lift system is the total number of passengers that it can transport in a period of 5 min during the up peak condition with a specified average car loading.

Obviously, meeting the high instantaneous demand would require large and expensive system. Thus a compromise is necessary, where the intending passengers are required to wait a reasonable time for service during peak demand periods. A period of 5 min for the handling capacity has achieved general acceptance. Thus if it is possible to equate the passenger demand as expressed by 5 min percentage peak arrival rate with the handling capacity of a system, then a suitable configuration could be designed.

4.2.4 Quality of Service :

The quality of service on the other hand is generally measured by the passenger waiting time at the various floors. The following shall be the guiding factor for determining this aspect:

Quality of Service or Acceptable Interval

20 to 25 seconds	Excellent
30 to 35 seconds	Good
36 to 40 seconds	Fair
45 seconds	Poor
Over 45 seconds	Unsatisfactory

Note: For residential buildings longer intervals may be acceptable.

4.2.5 Waiting time:

The theoretical average wait of all persons is one half the interval, the interval being the theoretical longest wait for any person. For example, if three elevators are provided and the average round trip time is 90 sec, the average interval for that period is 30 sec. Some people will get the service immediately with 0 sec wait and few may wait for the full 30 sec. The average waiting time in that case would be 15 sec. However, some times, some people may delay the elevator or the elevator may be filled and may bypass landing calls etc. Considering these conditions, the average wait time could be 55 to 60 % of the interval.

4.2.6 Traffic Peaks:

The maximum traffic flow during the morning peak period is usually considered as a measure of the vertical transportation requirement in an office building. The employees of all offices are subject to discipline and are required to be at their place in time. Consequently, the incoming traffic flow is extremely high and the arrival time is over a short period.

Sometimes it becomes necessary to reduce the maximum traffic flow by staggering the arrival of the employees so that different groups arrive at different times. This reduces the peak and also the requirement of lifts. However, many organizations may object to staggering and prefer to have all the employees arrive at the same time since it is claimed that staggering will affect the proper co-ordination of business.

4.2.6.1 Up-Peak Traffic:

An up-peak traffic condition exists when the dominant or only traffic flow is in upward direction with all or majority of passengers entering the lift system at the main terminal of the building.(The main floor is the arrival floor, that is the building entrance floor). Up peak occurs in considerable strength in the morning when lift passengers enter a building with intent on traveling to destinations on the upper floors of the building. The up peak generally occurs from the employers requiring their employees to arrive at work at a specified time. It is found that if a lift system can cope with the morning up peak, then it will cope up with other patterns of traffic, such as down peak and inter floor traffic. The traffic pattern is idealized by designers in terms of a 5min peak rate taken as percentage of the building population.

4.2.6.2 The up-peak arrival percentage:

is the number of passengers who arrive at the main terminal of a building for transportation to the upper floors over the worst 5 min period expressed as a percentage of the total building population.

A lift system is expected to respond to the peak demand in such a way as to quickly and efficiently transport passengers to their respective destinations without excessive passenger waiting time occurrence or formation of queues. This implies that the handling capacity of the lift system should be sufficient to carry all those passengers demanding service.

4.2.6.3 Down-peak traffic:

A down peak traffic condition exists when the dominant traffic flow is in a downward direction with majority of the passengers leaving the cars at the main floor of the lobby. To some extent, down peak is the reverse of the morning up peak occurrence at the end of the working day. It has been observed that the down peak is more intense than the morning up peak up to 50% higher demands and with durations of up to 10 min.

4.2.6.4 Two way and four way traffic:

A two way traffic condition occurs when the dominant traffic flow is to and from one specific floor, which is not the main terminal.

A four way traffic condition exists when the dominant traffic flows are to and from two specific floors, one of which may be the terminal floor.

They can arise from the presence of a refreshment floor which at certain times of the day attracts a significant number of stops and calls.

4.2.6.5 Inter floor traffic:

Random inter floor traffic can be said to exist when no discernable pattern of calls can be detected. Inter floor traffic is caused by the normal circulation of people around a building during the course of their business. Sometimes this traffic is called balanced two way traffic as it involves both up and down trips and it is balanced because passengers return to their original floors after moving about the building.

4.2.7 Speed:

It is dependent upon the quantity of service required and the quality of service desired. Therefore, no set formulae for indicating the speed can be given. However, the following general recommendations are made:

No of Floors	Speed (mps)*
4 to 5	0.5 to 0.75
6 to 12	0.75 to 1.5
13 to 20	Above 1.5

* Meters per second

4.2.8 Layout:

The shape and size of the passenger lift car bears a distinct relation to its efficiency as a medium of traffic handling. A study of most suitable proportions for these lifts reveal that the width of the well entrance is, in reality, the basic element in the determination of the best proportions. In other words, the width of the car is determined by the width of the entrance, and the depth of the car is regulated by the loading per square metre permissible under this standard. Center opening doors are the most practicable and the most efficient entrance units for passenger lifts.

4.2.9 Determination of Transportation or Handling Capacity during the Morning Peak :

4.2.9.1 The handling capacity is calculated by the formula:

$$H = 300 * Q * 100 / T * P$$

Where 300 = 5 minutes interval

H = handling capacity as the percentage of the peak population handled during 5 min period Q = average number of passengers carried in a car

T = Waiting interval and

P = total population to be handled during morning peak period (it is related to the area served by a particular bank of lifts)

The value of 'Q' depends on the dimensions of the car. It may be noted that the car is not loaded always to its maximum capacity during each trip and, therefore, for calculating 'H' the value of 'Q' is taken as 80 percent of the maximum carrying capacity of the car.

The waiting interval is calculated by the formula:

$$T \quad = \quad RTT / N$$

where

$$N \quad = \quad \text{No of lifts, and}$$

RTT = round trip time. That is, the average time required by each lift in taking one full load of passengers from ground floor, discharging them in various upper floors and coming back to ground floor for taking fresh passengers for next trip.

4.2.10 The round trip time (RTT):

The round trip time is the time in seconds for a single car trip around a building from the time the car doors open at the main terminal until the doors reopen when the car has returned to the main terminal floor after its trip around the building.

A round trip time should not usually exceed two to three minutes as the majority of this time can represent the journey time for some passengers with destination at top of the building, which is not desirable.

RTT is the sum of the time required in the following process:

a) Entry of the passengers on the ground floor.

b) Entry or Exit of the passengers on each landing floor.

c) Door closing time before each starting operation.

d) Door opening time on each passenger exit operation.

e) Number of Probable stops

f) Acceleration periods.

g) Stopping and leveling periods.

h) Periods of full rated speeds between stops going up, and

i) Periods of full rated speeds between stops going down.

It is observed that the handling capacity is inversely proportional to waiting interval which in turn is proportional to RTT. By reducing the RTT of a lift from 120 to 100 seconds its handling capacity increases by 20 percent.

The round trip time can be decreased not only by increasing the speed of the lift but also by improving the design of the equipment related to opening and closing of the landing and car doors, acceleration, deceleration, leveling and passenger movement. These factors are discussed below:

a) The most important factor in shortening the time consumed between the entry and the exit of the passengers to the lift car is the correct design of the doors and the proper car width. For comfortable entry and exit for passengers it has been found that most suitable door width is 1000 mm and that of car width is 2000 mm.

b) The utilization of center opening doors has been a definite factor in improving passenger transfer time, since when using this type of door the passengers, as a general rule, begin to move before the doors have been completely opened. On the other hand, with a side opening door the passengers tend to wait until the door has completely opened before moving.

The utilization of center opening doors also favors the door opening and closing time periods. Given the same door speed, the center opening door is much faster than the side opening type. It is beyond doubt that the center opening door represents an increase in transportation capacity in the operation of a lift.

4.2.10.1 An example illustrating the use of the above consideration is given below:

Net usable area per floor =950 sqm

No of landings including ground = 15

Assuming a population density = 9.5 sqm per person.

Probable population in 14 upper floors=

14 * 950 / 9.5

Total population to be handled during peak

P = 1400 persons.

If the calculated RTT Taking 20 passengers lift with 2.5 mps = 165 sec

Taking No of lifts N = 4

Waiting Interval T = RTT / N = 165 / 4 = 41 s

Waiting time = T/2 = 20.5 Sec

Avg No of persons carried in car in each trip

$$Q = 20 * 0.8 = 16$$

$$H = 300 * Q * 100 / T * P$$

$$= 300 * 16 * 100 / 41 * 1400 = 8.3 \%$$

Taking No of lifts N = 6

$$T = RTT / N = 165 / 6 = 27.6 \text{ s}$$

$$H = 300 * 16 * 100 / 27.6 * 1400 = 12 \%$$

Interval T = 27.6 s

Waiting time = T/2 = 13.8 s

4.3 Quiet Operation of Lifts :

Every precaution should be taken with passenger lifts to ensure quiet operation of the lift doors and machinery. The insulating of the lift machine and any motor generator from the floor by rubber cushions , or by a precast concrete slab with rubber cushions prevents transmission of most of the noise. Some recommendations , useful in this connection are given in IS 1950.

4.4 Position of Machine Rooms :

It will be noted that all lifts conforming to IS 14665 (Part 3 / Sec 1), have machine rooms immediately over the lift well, and this should be arranged whenever possible without restricting the overhead distance required for normal safety precautions.

Alternate machine positions should only be considered when there are special reasons justifying the additional cost, such as head room restrictions imposed by the planning authority for lifts serving the top floor.

It is desirable that emergency exit may be provided in case of large machine rooms having four or more elevators.

Two basic considerations, namely, the quantity of service and the quality of service desired, determine the type of lifts to be provided in a particular building. Quantity of service gives the passenger handling capacity of the lifts during the peak periods and the quality of service is measured in terms of waiting time of passengers at various floors. Both these basic factors require proper study into the character of the building, extent and duration of peak periods, frequency of service required, type and method of control, type of landing doors, etc.

For instance, they can with advantage be used to compare the capabilities of lifts in a bank with different loads and speeds provided the same set of factors is used for all cases.

4.5   RTT calculation Assumptions Leading to Traffic Analysis:

4.5.1 Entry & Exit of the passengers on the ground floor:

Usually the elevators are designed to wait for a fixed time of typically 8 seconds at the lobby for loading and unloading the passengers. It is found that typically an average of 0.8 seconds is required for transfer of passengers at the lobby. This means that up to 10 passengers can be loaded and unloaded within the fixed time of 8 seconds. If there are more passengers, 0.8 seconds should be added for each passenger.

4.5.2 Entry or Exit of the passengers on each landing floor:

The lifts are typically designed for a dwell time of 2 sec when the lift halts at a particular landing, initiated either by a car call. Similarly the dwell time due to landing call is 4 Seconds..

4.5.3 Door closing & Door Opening time at each landing:

Typically the elevators are designed to have a shorter door opening time for the passengers to exit the lift quickly. The door closing time is kept more to avoid the doors banging on the passengers who are entering the lift.

The table below provides some assumptions on typical door open and close times. This may vary from manufacturer to manufacturer. It will be a good idea to get these values from the manufacturers before doing further calculations.

Door Type	Width in mm	Open t_o (sec)a	Close t_c (sec)	Total (sec)b
Two Speed	900	2.1	3.3	5.9
Center opening	900	1.5	2.1	4.1
Two Speed	1100	2.4	3.7	6.6
Center opening	1100	1.7	2.4	4.6

Table 4.1 Typical Door Operating Times
a) Door open time can be reduced by 1 sec in case of advanced door opening
b).Once the door is closed, it takes some time to ensure that the door is locked and for the elevator motor to build up torque to commence run. This

delay is estimated as 0.5 sec and added in the above door operating table.

4.5.4 Probable number of stops (S) :

a) The number of people entering the lobby greatly influences the number of stops the elevator makes. The number of floors the elevator serves influences the number of stops an elevator is expected to make with a given passenger load. For example, an elevator that serves 8 floors and carries 8 passengers is most unlikely to stop at all the 8 floors.

b) The nominal population in each floor also has direct influence on the number of stops made per trip. If one floor has a population of 100 and the other 10, the tendency for stopping at the floor with a population of 100 is much more.

c) When there is an incoming traffic peak, the elevator makes predominantly up car stops and returns to the lobby with very small probability of in between stops. When everyone wants to leave the building after office hours, a different pattern of stops is likely.

d) When persons are traveling in between floors, each person makes two stops, one for boarding and the other for leaving. Under extremely busy condition with inadequate elevators, it is possible for an elevator to make every stop up and every stop down, resulting in intolerably high waiting times.

e) In estimating the probable number of stops per trip, we can reasonably be certain that in every elevator trip the elevator will stop in upper floors in proportion to the number of people in the car and the number of floors that elevator serves. George Strakosh in his book titled Vertical Transportation Elevators and Escalators has provided a statistical calculation of the probable number of passengers leaving the elevator at a given floor at the same time with the following formula:

f) Probable number of stops per trip $= S - S \left[\frac{S-1}{S}\right]^{p}$

S = Total number of stops above lobby
p = No of passengers carried on each trip

A simple calculation with the above formula for floors above lobby upto 20 floors and total capacity of 20 Passengers leave us with the following table:

		No of passengers per trip									
		2	4	6	8	10	12	14	16	18	20
	20	2	3.7	5.3	6.7	8	9.2	10.3	11.2	12.1	12.8
	18	2	3.7	5.2	6.6	7.8	8.9	9.9	10.8	11.6	12.3
	16	2	3.6	5.1	6.5	7.6	8.6	9.5	10.3	11	11.6
No of Floors above lobby	14	2	3.6	5	6.3	7.3	8.3	9	9.7	10.3	10.8
	12	2	3.5	4.9	6	7	7.8	8.5	9	9.5	9.9
	10	2	3.4	4.7	5.8	6.6	7.2	7.7	8.2	8.5	8.8
	8	2	3.3	4.4	5.3	5.9	6.4	6.8	7	7.3	7.5
	6	2	3.1	4	4.6	5	5.3	5.5	5.7	5.8	5.8
	4	2	2.7	3.3	3.6	3.8	3.9	3.9	4	4	4
	2	1.75	2	2	2	2	2	2	2	2	2

Table 4.2 Probable Stops

Above Equation is the one used in planning elevators for buildings when little is known about future population and each floor is assumed to have equal population. Applying the formula, Table 4.2 shows probable stops value for approximate numbers of passengers per trip and total number of upper floors served by an elevator.

4.5.5 Running Time:

Running time includes:4.5.4 Number of Probable stops

i)   Acceleration & Deceleration periods.

ii)   Starting, Stopping and leveling periods.

iii)   Periods of full rated speeds between stops

Assumptions:

(i) Acc / Dec Rates

0.6 m / sec / sec for 0.7 mps speed
1.0 m/sec/sec for 1 mps speed
1.1 m/sec/sec for 1.5 mps speed
1.2 m/sec/sec for 1.75 mps speed
1.25 m/sec/sec for 2 mps and speed
1.35 m/sec/sec for 2.5 mps speed
(ii) Starting + Stopping delay = 0.5 sec (for DC injection at Zero speed)

(iii) S curve delays during acceleration + Deceleration = 1 Sec

Using the Newton's law v = u+at and v*v + u*u = 2as

Where v = final velocity, u= initial velocity, a = Acc /Dec rate & s = distance travelled, the following table gives time for various speeds and different acceleration / deceleration rates

	Speed					
	0.7mps @ 0.6 m/s/s	1 mps @ 1 m/s/s	1.5 mps @ 1.1 m/s/s	1.75 Mps @ 1.2 m/s/s	2 mps @ 1.25 m/s/s	2.5 mps @ 1.35 m/s/s
Time in s	2.3	2.0	2.7	2.9	3.2	3.7
Dist in m	0.8	1.0	2.0	2.6	3.2	4.6

Table 4.3 Time and distance during acc / dec

During time other than the acceleration and deceleration, the lift will run at rated speed.

Time during rated speed = Distance travelled / Speed.

With the running time assumptions, above table 4.3 and time during rated speed, time for floor to floor can be calculated.

The table below provides the total run time

Speed	Floor Heights							
	3	3.5	4	5	6	7	11	For each meter
0.7	6.9	7.7	8.4	9.8	11.2	12.7	18.4	1.4
1.0	5.5	5	6.5	7.5	8.5	9.5	13.5	1
1.5	4.9	5.2	5.5	6.2	6.9	7.5	10.2	0.67
1.75	4.6	4.9	5.2	5.8	6.3	6.9	9.2	0.57
2	4.6	4.9	5.1	5.6	6.1	6.6	8.6	0.5
2.5	4.6	4.9	5.1	5.5	5.9	6.3	7.9	0.4

Table 4.4 Run Time

In addition to loading and unloading and door opening and closing, the other element in the round trip time of the elevator is the running time. Part of the running time is spent in accelerating to rated speed and decelerating it to stop. Typical floor to floor traveling time for elevators of various speeds and with typical acceleration and deceleration rates is given in table 4.4

The above table indicates that increase in speed helps to reduce the time only when the distance travelled is more.

Sample calculation for 1.5 mps speed and travel of 6 m.

From table 4.3, time for acceleration & deceleration is 2.7s and distance travelled during acceleration/deceleration is 2.0m.

Time during rated full speed = (6-2) / 1.5 = 2.7s

Total time= Start/Stop delay + S curve delay+ Acc/Dec time + Rated speed run time

Total time = 0.5 + 1 + 2.7 + 2.7 = 6.9s

4.5.5 TRAFFIC CALCULATION - EXAMPLE

Assumptions	Example 1	Example 2	Example 3	Example 4 w/ Ad door op
No of Passengers P	16	20	20	20
Speed mps	2.5	2	1.5	1.5
No of Floors	11	11	11	11
Floor to Floor Height met	3	3	3	3
Door C/O mm	900	1100	1100	1100

SOLUTION:

1	No of Probable Stops	8.2	8.8	8.8	8.8	Refer Table 4.2
2	Time to load passengers at the lobby - Sec	14.4	16	16	16	4.5.1
3	Transfer time at	16.4	17.6	17.6	17.6	2 sec at each probable stop.

	upper floors Sec					
4	Time to open and close door - Sec	4.1	4.6	4.6	3.6	Refer table 4.1
5	Total Door operation time	37.7	45	45	35.3	(Probable stops + Ground Floor) Multiply by S # 4
6	Total Time Spent at floors Sec	68.5	78.6	78.6	68.9	*Add S# 2, 3 & 5*
7	Apply10 % inefffor loading + door ineff	75.4	86.5	86.5	75.8	Multiply S No 6 by 1.1
8	Total distance travel - met	30	30	30	30	Floor to Floor dist=3m Total no of floors=10 Total distance=30m
9	Dist to run each floor in UP dir	3.7	3.4	3.4	3.4	Divide S No 8 by S No 1
10	Time to	3.4	3.3	3.63	3.6	Refer table 4.4

	run @ rated speed - Sec	5			3	
11	Total time to run UP- Sec	28.3	29.0	32	32	Multiply S No 1 and 10
12	Time to run from top floor to lobby	13.9	16.6	21.4	21.4	Refer Table 4.4
13 *	RTT*	118	132	140	129	Add S # 7, 11and 12
14	5Min capa	32.5	36	34	37	=300*0.8*P/RTT
15	No of lifts reqd	6	6	6	6	Assuming 25 sec interval, RTT/25
16	5 min handling capacity	195	216	204	222	Multiply S No 14 and 15
17	Handling capacity %	16.3	18	17	18.5	*Assuming 1200 Persons in the building, S# 15/1200 as %*
18	Revised interval- sec	20	22	23.3	21.5	RTT/6

Summed up results of Examples 2 & 3:

	With 2	With 1.5	With 1.5

	m/s Speed	m/s Speed	m/s Speed &Adv Door opening
No of Cars	6	6	6
Interval	22	23.3	21.5
5Min Handling Capacity	216	204	222
HC %	18	17	18.5

It may be noticed that with advanced door opening, the performance of elevator with 1.5 m/s is comparable to 2 m/s speed, with all the other specifications remaining the same, as the round trip time has improved with reduction in door open times.

*For a single car, RTT in S No 13 = Interval.

CHAPTER 5

CAR, CAR FRAME, PLATFORM, ROPES & TRACTION

5.1 Car / Cab and Its Construction :

We know that the carriage in which a passenger travels up and down is known as a Car or Cab.

The car enclosure is usually made of sheet metal members and bolted together to form a cage. In general the car is made either in square or rectangular shape. Very often the residential buildings decide to have a rectangular shaped deep car in order to house a stretcher which may be required to carry sick persons, at times.. There are round, Hexagonal, Pentagonal etc shaped cars also which are custom designed especially for use in shopping malls or public places to improve the aesthetics of the building. These panoramic cars generally are designed with glass enclosures. The Bureau of Indian standards provides recommended dimensions of the car inside, the Lift well and the door opening sizes.

5.2 What is a car frame >>

A car-frame is a rigid rectangular frame consisting of a cross head, uprights, and plank see the figure below to get a basic understanding of the car frame assembly. The car frame must be constructed to withstand the weight of the car, passengers, platform, safety devices, doors and door operator and all other loads pertaining to the cab system. It should also be designed to withstand safety operation under maximum load conditions described above.

- The car-frame generally comprises a safety plank, two uprights and a crosshead.

- The Cross head, which forms the top of the car frame, consists of structural members generally channel shaped.

- Uprights, are the vertical structural members at the sides of the car frame.

- Plank is a structural member similar to the cross head forming the bottom of the car frame.

- The roller guides or guide shoes are located at the four corners of the car frame. The guides help to guide through the rails. As per BIS, where the platform is directly supported by the plank or sound isolation frame, the vertical center distance between the top and bottom guide shoes should be more than the distance between the guide rail brackets.

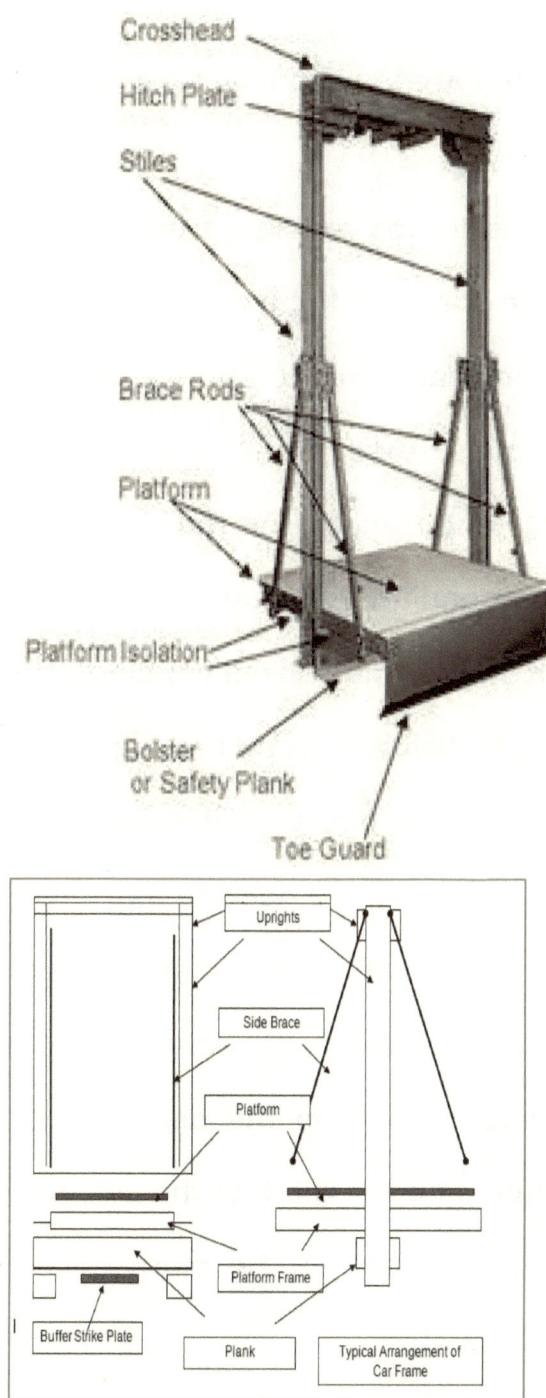

Crosshead

Hitch Plate

Stiles

Brace Rods

Platform

Platform Isolation

Bolster
or Safety Plank

Toe Guard

Uprights

Side Brace

Platform

Platform Frame

Buffer Strike Plate

Plank

Typical Arrangement of
Car Frame

133

5.2.1 What are the important considerations in a car frame >>

The forces acting on the car frame members are quite complex. If the center of gravity of the rated load in the car coincide with the center of load on the hoist ropes, then there would be tension in the uprights and bending on the cross head and plank. But this condition will not occur as the passengers are not static. This causes bending in the upright and twisting in the cross head and plank.

Since the exact position of the load at a given time is unknown we need to make few assumptions when the elevator is being loaded and during running. Any load that is entering the car can not be more eccentric from the center of the car to the maximum opening width of the car. Similarly, the car once loaded can not be more eccentric from the centre of the car to the physical size the cab wall will permit.

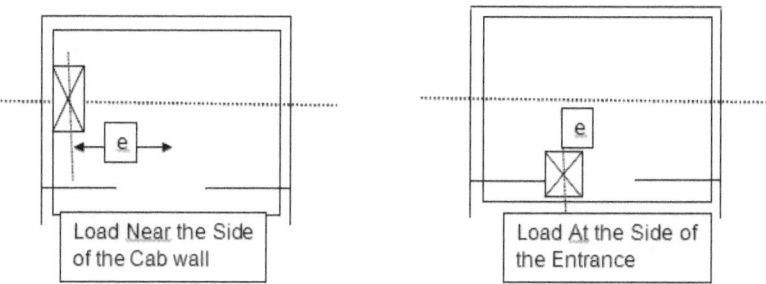

Load Near the Side of the Cab wall

Load At the Side of the Entrance

The two worst case conditions which give us the starting point for making the calculations. In both the above cases, the car frame is subjected to an overturning moment, the magnitude of which is equal to the load times the eccentricity. The guide shoes resist this overturning effect and impart pressure on the guide rails. This causes the upright to deflect and cause the platform to sag on one side. Generally the cross head and the plank do not pose major design issues but the car frame uprights pose a few design issues. The uprights are designed to occupy less space in order to maximize the car area. In doing so, one has to sacrifice the strength in the uprights. When the rated loads are high causing large car frame bending moments and the available hoist way is small, double uprights are used.

The point of application of the loads with reference to the entrances both while loading and unloading as well as the position of the load in the car while the elevator is running greatly influence the size of the car frame members. In the absence of loading conditions extreme conditions must be assumed for the calculations resulting in much heavier construction than actually required.

In addition, during safety application the car frame must withstand the forces impressed when the safeties grip the rails. Owing to the tolerance in dimensions of the two rails and the difference in lubrication of the two rails, the retarding force on each of the rails will not be the same. The difference in forces cause over turning moment and the over turning moment causes bending of the uprights.

The other extreme condition is at the time of buffer engagement. The buffer is a device located in the pit at the bottom of the hoistway. The plank channels are subjected to severe bending load. The uprights undergo compression and bending depending upon the position of the load in the car at that moment.

5.3 What is a Platform?

Elevator car platform is defined as that structure which forms the floor of the car and directly supports the load. The modern elevators utilize an all steel platform construction which is lighter and efficient than its wooden predecessors. In the all steel version, the design might consist of several sections welded together wherein the stringers and floor plates are the modules. The platform or platform frame is attached to the upper side of the safety plank and uprights of the car-frame. The platform provides the mounting surface for the cab enclosure. In most cases, the car is isolated from the platform using rubber materials. This rubber isolation provides noise and sound isolation to the car. The compression of the rubber is also used to detect the load in the car by fixing micro switches beneath the rubber isolation. Due to pressure of the rubber materials, the Micro switch which is kept below gets operated and this is used to sense the load in the car. This rubber isolation of the car from the car frame also prevents vibration passed on to the car.

5.4 Roping methods:

For 1:1 roping, the crosshead members provide mounting for the hitch plate which is positioned under the crosshead

For 2:1 roping, the crosshead members provide mounting for the hoisting rope sheave, positioned above, within or below the crosshead.

For under slung car frames the 2:1 sheaves are provided under the safety plank. .

Car frame for 1:1 roping and that for 2:1 roping with a single sheave are shown below.

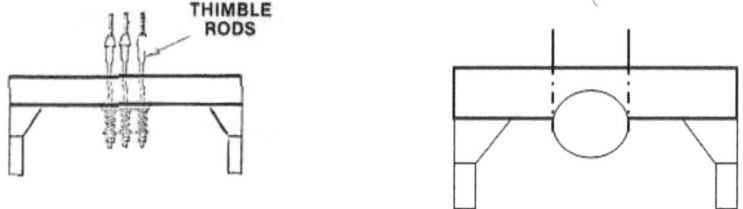

5.4.1 What is 1:1 Roped System ?

Shown below is the basic arrangement of elevator car and counterweight, where the car and counter weight hang on either side of the driving sheave of the machine

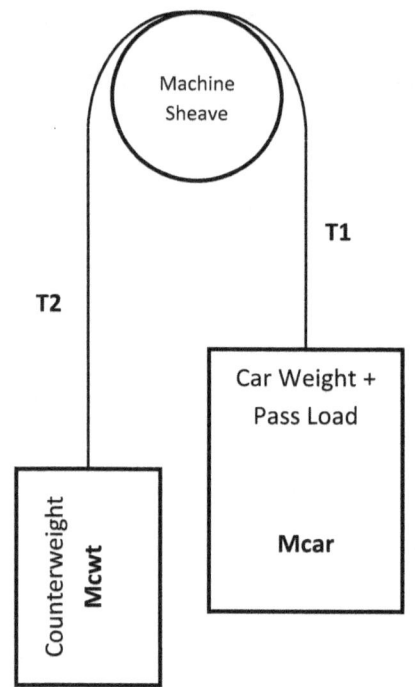

Tension in ropes/belts on car & counterweight side can be calculated by using the following formulas:

$T1 = (Mcar + P)* g$

$Mcwt = Mcar + P/2$

$T2 = (Mcwt)*g$

where g is gravity constant

P=Passenger load

SSL (Sheave Shaft Load)= T1 + T2

Rope masses are not included in tension calculation in this example

Roped System 1:1

RATINGS OF MACHINE AND MOTOR:

Let us assume a duty load of 10 Passengers with a car weight of 1000Kg

T1 = (1000 + 680) Kg as each passenger weight is assumed as 68Kg

T2 = (1000+ 340)

The maximum unbalanced load = T1-T2 = 340Kg

Please note that the load in the cwt is fixed whereas the load in the car is variable. Assuming no passenger in the car

T1=1000Kg ; T2 = 1340 and T1-T2 = -340 Kg.

This load "F" has to be moved UP or DOWN.

When the car is empty, UP movement of car does not require any power as CWT is heavy. Similarly when there is full load in the car, DOWN movement of the car does not require any power as the car is heavy.

Assuming a sheave diameter of 610mm, the radius of the sheave is 305mm.

The Load torque required is F * R

F = 340Kg; R = 0.305m;

Hence Load Torque = 103.7 KgMtr

Motor Torque = Load Torque / Gear Ratio * Efficiency

Assuming Gear Ratio = 46; Efficiency = 50%,

Efficiency here is a combination of motor, gear and frictional losses in the rails.

Motor Torque = 103.7/ 46*0.5 = 4.5 KgMtr

Ie 45NMtr (10NMtr = 1 KgMtr)

Considering the voltage and frequency fluctuations, overloading etc one rating higher than the calculated rating is taken.

Motor Power = (Load in Kg * Speed) / 75 * Efficiency

$$= (340 * 1)/(75*0.5) = 9HP$$

(The factor 75 is a constant calculated from fundamentals).

Final selection of motor rating is determined based on torque. To be on safer side, higher HP and higher torque over the calculated value is chosen.

5.4.2 WHAT IS ROPED SYSTEM - 2:1?

The effect of 2:1 roping is to essentially double the lifting force related to the rope tensions, and another effect is that elevator car speed is a half of machine sheave speed. 2:1 roping is widely used in elevator systems.

Flexible cables are used to bring electrical connections from the machine room to the car.. Since the wires travel with the car, they are called traveling cables.

Some ropes are used to compensate for the weight of the hoist ropes in order to equalize the loads on the car and counterweight sides at different

positions of the car in the hoist way. These are called compensation cables or compensation ropes.

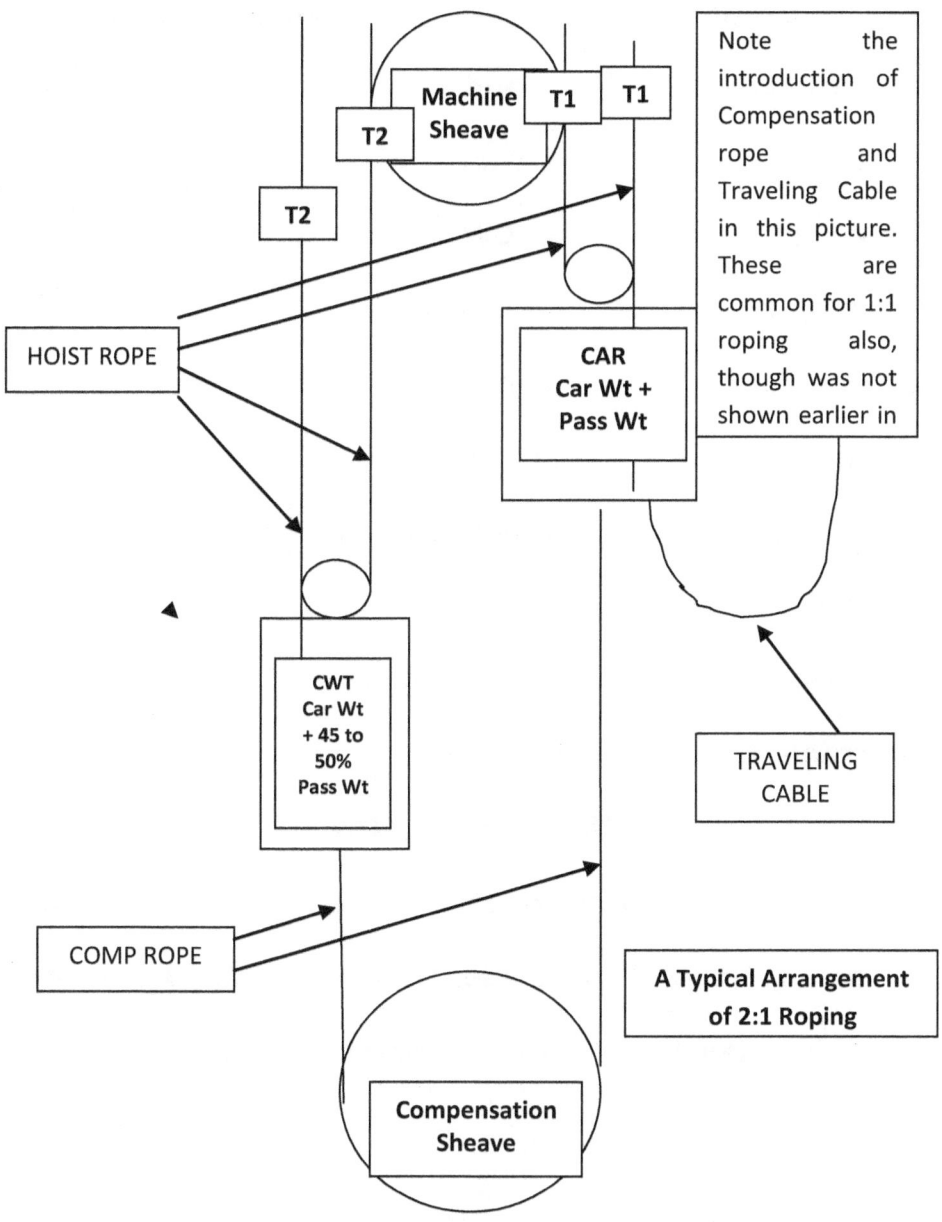

Note the introduction of Compensation rope and Traveling Cable in this picture. These are common for 1:1 roping also, though was not shown earlier in

Machine Sheave
T1
T1
T2
T2
HOIST ROPE
CAR
Car Wt + Pass Wt
CWT
Car Wt + 45 to 50% Pass Wt
TRAVELING CABLE
COMP ROPE
Compensation Sheave

A Typical Arrangement of 2:1 Roping

5.5 WHAT ARE DEFLECTORS?

In actual case it would not be possible to hang the car on either side of the drive sheave due to various sizes of car and main sheave dia. In such cases deflector sheaves are used to obtain rope drops necessary for CWT suspension, and also to support the car as close as possible to the car's center of gravity. Arrangement with deflector sheave has two main effects:

Sheave shaft load is reduced when compared to 180 degrees wrap

Wrap angle is reduced, which reduces available traction

5.5.1. SINGLE WRAP TRACTION:

Single wrap traction is a relatively simple arrangement that provides a rope drop and sufficient wrap angle to deliver a required traction.

Arrangement shown below is a 1:1 roping arrangement with deflector sheave.

In the example shown below, the diameter of main sheave and the deflector sheave are the same.

CWT

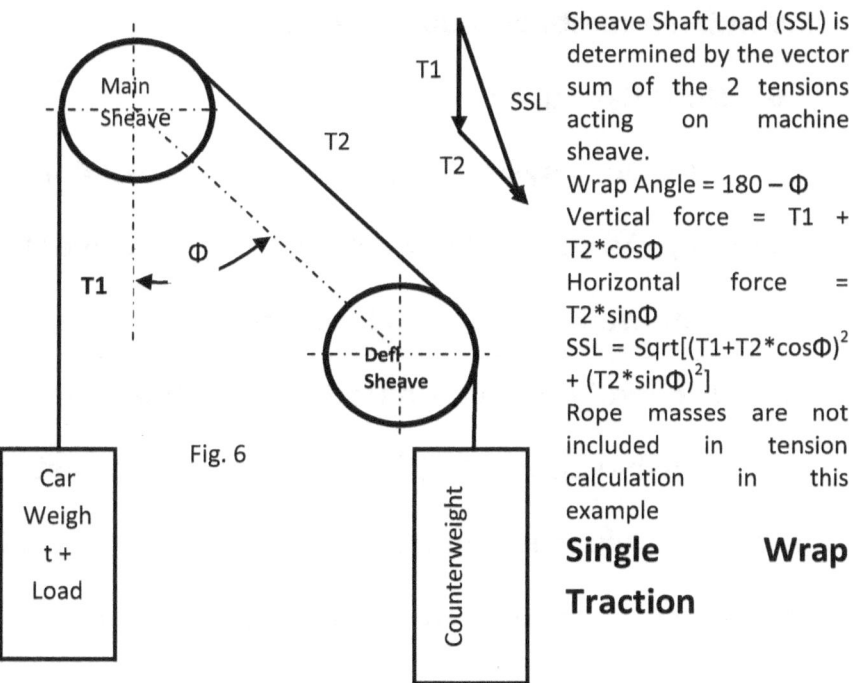

Sheave Shaft Load (SSL) is determined by the vector sum of the 2 tensions acting on machine sheave.

Wrap Angle = 180 − Φ

Vertical force = T1 + T2*cosΦ

Horizontal force = T2*sinΦ

SSL = Sqrt[(T1+T2*cosΦ)^2 + (T2*sinΦ)^2]

Rope masses are not included in tension calculation in this example

Single Wrap Traction

Fig. 6

CWT= Car Weight + 45 to 50% of Passenger weight.

Assuming T1=T2=100 and φ =30; Cos 30= 0.86 and

Sin 30 = 0.5

SSL = $\sqrt{((100 + 86)^2 + 50^2)}$ = 192.6 .

The above equation shows that the SSL reduces with deflector.

But for the purposes of calculations we always consider SSL = T1 + T2. This is a safer calculation.

5.6 Sub- Assemblies mounted on car:

5.6.1 Door Operator:

The Door system is also a part of the total car assembly. As per Indian Standard, the movement of car is allowed only with closed door. .(Except in special conditions such as advanced door opening or Re-leveling operation)The door system consists of a car door which is opened or closed either manually or automatic. The early designs of manual doors consisted of Collapsible gates. Considering passenger safety, only imperforated doors are currently used for manual door system. On closure of car door, a car gate switch indicates to the controller logic that the car door is closed and the lift can run.

5.6.2 Are there doors in each landing?

Each landing is provided with a landing door. These doors remain mechanically latched so that when the lift is not in that floor, no one can inadvertently open the landing door and inadvertently fall into the hoist-way. The car gate gets mechanically engaged with the landing gate only while at the corresponding floor level. Due to this mechanical engagement between the car door and the landing door, the car gate moves the landing gate along with it. When the car gate is fully closed, the landing gate also gets fully closed and both of them through their own electrical contacts provide independent signals to the elevator controller that both the car door and landing doors are closed. When the car leaves the floor, a lock gets mechanically engaged with that corresponding landing door and hence the landing door remains closed & latched until the car once again reaches that floor to stop and gets mechanically engaged. All the landing doors remain mechanically locked so that no landing door can be opened when the car is not in that location. When the car reaches the landing, the car door gets mechanically engaged with the corresponding landing door and hence the landing door can be either opened or closed by the motion of the car door.

5.6.3 Safety conditions to be ensured for the car to run:

Following safety requirement is ensured before an elevator car moves out of a landing:

a. Both the car and all the Landing doors are fully closed. This condition is ensured by sensing electrical signal through a car gate switch to confirm that the car gate is fully closed. All the landing gate switches are electrically wired serially so that inadvertent opening of any one of the landing door will immediately stop the lift and will not allow it to run. It shall not be possible to start the car in motion unless the car door and all the landing doors are fully closed.

b. All the landing entrances to be protected by doors which will remain locked and cannot be opened from landings except in case of emergency and by authorized persons only. It must be possible to open the landing gate only when the car is present and stopped at that landing. In case of manually operated doors, this safety requirement is achieved by energizing an electro magnet known as "Retiring Cam" which is mounted on the car top. The retiring cam essentially consists of an electro magnet and a cam which actuates an electro mechanical lock in the landing and mechanically unlocks the landing door. When the roller of the landing door electro mechanical lock is pressed by the retiring cam plates, the landing door will be unlocked.

5.6.4 Power Operated Door:

A door which is opened or closed by motive power other than hand is called as Automatic door. An electric motor is used to close or open the car door either using mechanical levers or by using tapes and pulleys. The motor used is either DC, induction or Permanent magnet type. Present day designs have PM motor whose speed is precisely controlled by Variable voltage variable frequency (VVVF) drive. Both the door drive and the door motor are mounted on the car top. This drive gets commands from elevator controller (either directly or through car top printed circuit board for door opening, Closing, Door Zone etc.

The automatic doors are designed such that the opening of the door can occur only at the landing within a specific zone called the Leveling Zone while the car is at rest. This is accomplished by sending a door zone signal to the elevator controller and the VF drive for the door motor which enables opening of the door.

As mentioned earlier, the car door and Landing door are mechanically coupled so that they open and close together. The coupling is done by cams fitted on the car and Rollers on the Landing door. The car can move from a floor only when the car and landing door electrical contacts are closed and the landing door is mechanically locked. This mechanical locking of landing door is essential to prevent possibility of opening a landing door when a car is not in that landing.

Movement of elevator without closing the doors can result in fatal accidents. Such a condition can arise due to failure of any of the components such as breakage of ropes or pulleys. A copy of the Failure mode effect analysis conducted by the manufacturing company can be obtained to ensure that the elevator controller will not get a wrong door closed signal under any failure condition.

5.6.4.1 Safety measures while using an auto door:

The power operated doors are designed to exert limited force so that they do not hit and injure a person.

This is accomplished in two ways:

a) The force inserted by a closing door is designed to be less than 150N. Some of the door systems of present designs re-open the door on sensing the pressure.

b) Obstruction to the door closing movement is sensed by operating a momentary pressure operating switch, which provides a door open command to the door drive system.

c) The obstruction is sensed electronically by infra-red rays mounted vertically along the car door so that the door can open on sensing obstruction even without actually touching the passenger. This arrangement is very commonly used in the elevators manufactured at present.

5.6.5 Few other assemblies are mounted on the car:

Electronic controls are used to smoothly open and close the doors on receipt of commands from the controller. On reaching the door open limit (Fully Open) or door close limit (Fully closed) the power to the door motor is brought to zero. The present day door operator controllers can learn the door position by counting the pulses of the encoder mounted on the door motor shaft. This helps to achieve smooth opening and closing of the doors.

Car is also fitted with fans, lights and emergency light and false ceiling in most of the cases.. The Fans are electronically controlled that they go off after a delay when the lift is idle. This helps to save the electrical energy and increase the life of the fan. A car is fitted with emergency light to provide light in the car at times of power outage. Over load announcement and indicators are also common features found in the car to inform overloaded condition to the passengers. The lift does not start under over loaded conditions. Some customers desire voice announcement of floor arrivals in one or multiple languages to support the disabled.

The most important assembly is "car top inspection box assembly". Often, the elevator mechanic is required to run the elevator from the car top for doing maintenance work. The Car top Inspection box assembly consists of STOP push button to stop the elevator, a rotary switch to turn to maintenance mode of operation, two push buttons UP and DN to decide the direction of travel and one common push button. Inspection speed should not exceed 0.7m/s. The service man is required to press a common button and a direction button together to move in a particular direction. The lift stops once any one of the two switches is not pressed. The inspection operation circuit and fixtures are required to be designed with utmost care to prevent any un-intentional movement of the car even in case of failure of any component. It is strongly advised that the elevator manufacturing companies conduct

" Failure Mode & Effect Analysis"   (FMEA) to prevent un intended movement of car even in case a failure of wiring or any one component. There has to be ways to detect a first failure. Failure of any component must be detected, bring the lift to stop (without trapping passengers) and create a break down so that the fault can get rectified before a second different failure occur. The designer should keep in mind to ensure elimination of elevator accidents because lift accidents are normally either fatal or very severe.

Provision is also made on the top of car to load the car with weights. Since many components are mounted on the car, the weights may tilt the balance in static condition. To balance the car under static conditions provision is made to load the car using weights.

Other important assemblies are the guide shoe and roller guides.

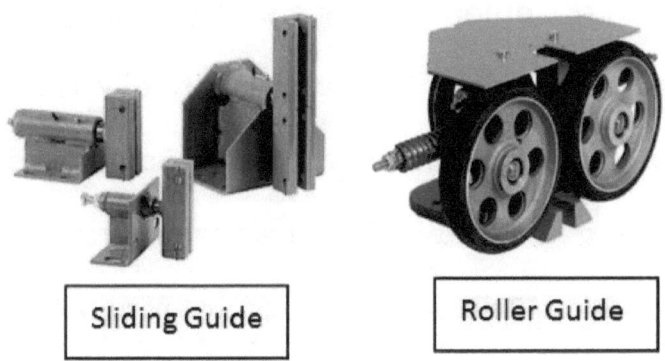

| Sliding Guide | Roller Guide |

Typically Sliding guides are used for speeds up to 2 mps and Roller guides are used for speeds above 2.5 mps. Roller guide comprise of spring loaded rollers which are in contact with the three sliding faces of the guide rail. The rollers are lined with rubber or polyurethane tires. Noise or vibrations are reduced and the ride quality is significantly improved since the rollers are mounted on ball bearings. For very heavy duty lifts there are designs with six rollers. Roller guides operate on dry rails and hence no oil or any lubricant need be used on rails.

The other major safety component is mechanical safety. These are normally fastened to the underside of the safety plank. Safeties may also be arranged to attach to the elevator structure in the top position. The safety levers are roped to a safety governor which is located in the machine room. When the speed increases beyond the set value the safety levers actuate the safeties which grip firmly with the guide rails and prevent any further mechanical motion. Safety blocks are the ultimate safety for elevators to prevent a free fall of elevator due to ropes cut. Safety in general gets activated only during downward motion. The operation of mechanical safety also sends electrical signal to the controller to cut off the power supply to the motor.

Instantaneous safeties are used for speeds not greater than 1 mps. Progressive safeties are used for speeds above 1mps.

5.6.6 What is a car operating panel (COP) :

The car operating panel is an assembly of push buttons, a display and other switches, located inside the car. The push buttons facilitate the passenger to register calls to go to desired floor. The buttons in the car operating panel consists of a contact, which when pressed signals the Car Top board which in turn communicates serially with the main controller kept in the machine room. The main controller acknowledges the call and communicates to the car top board which in turn lights the LED in the corresponding call button. This is how a light turns ON whenever we press a car call button. In earlier versions of elevators, the call button wires were taken directly to the controller in the machine room. The method of communicating with the master controller either serially or in parallel is based on the cost benefit analysis of running a number of wires to the machine room VS using few serial communication wires. Reducing the number of traveling cable wires also improves the failure rate. All the wires connecting a car to the machine room have to be wired using special traveling cables. This is required due to the number of bending operations the wire undergoes as the wires are connected to the car which moves continually UP and Down.

The other commonly used buttons in a COP are Alarm, Interphone, FAN switch etc The COP is also used for giving Door Open, Door Close commands. The COP has hands free intercom which facilitates intercommunication between the passenger inside the car and the elevator machine room or Security. The Alarm button is connected to a hooter which is kept on the car top. The Alarm has to be active even when power fails in order to get attention in case of entrapment inside the car. The interphone and Alarm are provided to help a trapped passenger inside a lift.

Typically the COP will have Floor Call buttons, Door Open Button, Door Close button, Alarm button, Auto / Attendant switch, Independent control, Interphone, Fan Switch, Interphone switch and position indication display.

Car Top Box: The car top box communicates with the master controller serially. The other electronic assembly which communicates serially with the car top box is the car call panel CCP. The CCP is mounted in the Car operating panel COP and all the car call buttons, Door Close, Door open and other buttons are hard wired to this CCP. The car top box also controls the door drive unit. The car top box sends "Door Close" or "Door Open" commands to the door drive.

5. 7 ROPES AND TRACTION:

A wire rope has three main elements

a) Wire

b) Strand

c) Core

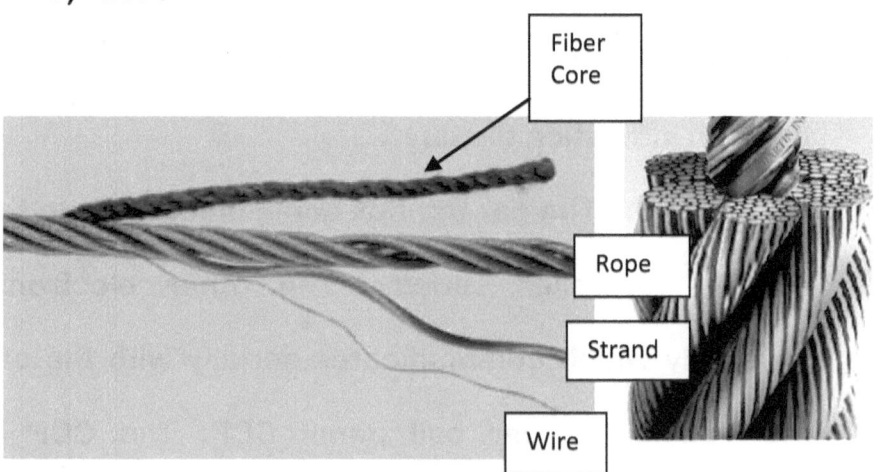

Source: Usha Martin

The basic component of a wire rope is a wire, which is made up of steel in various sizes. The number of wires in a strand depends on the usage of the wire rope. A defined number of wires are spun helically around a central wire, which is called a strand. A number of such strands are then helically spun around a core to form a wire rope. The way the wires are spun to form the strands and the way these strands are spun around the core determine the overall performance characteristics of the wire rope.

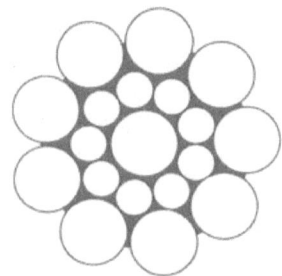
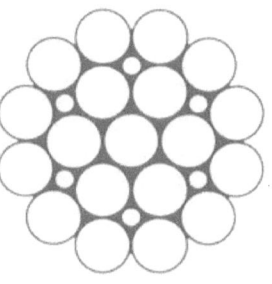

SEALE	**WARRINGTON**	**FILLER**
This arrangement has larger diameter wires in the outer layer and has an equal number of wires in the inner layer to provide flexibility.	This arrangement has Alternate arrangement of smaller and larger wires on the outer layer. Warrington has flexibility and abrasive properties.	Filler wires have smaller wires filling the empty spaces between the outer and innerlayer. This configuration offers better fatigue properties and good abrasion resistance.

Source: Usha Martin

The geometrical arrangement of wires in a strand is called its construction. The most common strand constructions are Seale, Warrington and Filler.

While ordering ropes, it is important to mention the following

163

- Nominal rope diameter (eg 10mm)
- Rope Construction (eg 8 X 19 W meaning 8 strands of 19 wires each Warrington arrangement)
- Type of core
- Tensile Grade
- Surface coating
- Lay – Type and direction.

5.7.1 Rope Selection:

The number of expected bending cycles a rope has to face is a key information before selection of a rope.

Foresee that a particular installation will face continuous use with high acceleration and deceleration.

The following parameters are used for rope selection:

- Code required Factor of Safety

- Actual Factor of Safety

- Number of ropes

- Static tension per rope

- Traction ratio

- Popular rope sizes are

  8mm,10mm,13mm and 16mm.

Ropes should be selected per the breaking loads provided by their manufacturers, and the Factor of Safety (FOS) required by the local codes. The most important aspect of rope selection is the verification of the factor of safety and must generally comply with a factor of safety = 10 for speeds up to 2 mps and greater for higher speeds.

The BIS specifies the factor of safety for suspension ropes In the case of the traction drive, the factor of safety shall be based on static contract load plus the weight of the lift car and accessories.

All the ropes at one installation must be from the same manufacturer and of same material, grade, construction and diameter preferably cut from the same reel.

The factor of safety of the wire rope can be calculated by the following formula:

$$f = \frac{S \times N \times K}{W}$$

K = Number of runs of the rope. For 2:1 roping this shall be 2

N = Number of ropes

S = Manufacturer's rated strength of one rope.

W = Maximum static load imposed on all the car ropes at its rated load on any position of car in the hoistway.

5.7.2 Example– Calculation of factor of safety

Elevator Specs		Manufacturer's Catalog	
Number of ropes (N)	3	Rope unit weight (kg/m)	0.35
Duty Load (	1088	10mm Rope	40,000

Kg) DL		braking force (N)	
Rise (m)	60	Trav Cable unit weight (Kg/m) tc	1.1
Overhead (m)	4.5	Comp Rope unit weight (Kg/m) tcomp	2.24
Pit depth (m)	1.6		
Dia of Ropes mm	10		
Roping	2:1		
CWT Over balance	45%		
Total Car weight (Kg) Carw	1300		

Calculations

Traveling cable weight Trw= R *tc*0.5	33	
Total Counter weight= Carw+0.45*DL+0.5Trw	1806	
Total Comp chain weight Compw= R* tcomp	134	
Total Rope length= 2* Pit depth + 2* Rise + 4* OH	141.2	
Weight of ropes= Rope length* unit weight*No of ropes	148	
Car on top level with 125% load= 1.25*DL+Trw+Compw+Carw	2827	
Car at bottom level with	2808	

125% loaded= Carw+1.25*DL+Rope weight		
Safety Factor= N*(Min braking force / Max weight on car side) Assuming N=2	21.3	
Acceptable limit as per BIS	12	
Even with N=2, it meets the Safety Factor. But less than 3 # 10mm ropes are not acceptable as per BIS Hence use 3#, 10mm Ropes.		

For a long rope life, ensure

- Lubrication

- Proper alignment of sheaves

- Ropes are not twisted during installation

- Equal adjustment of rope tension at the installation.

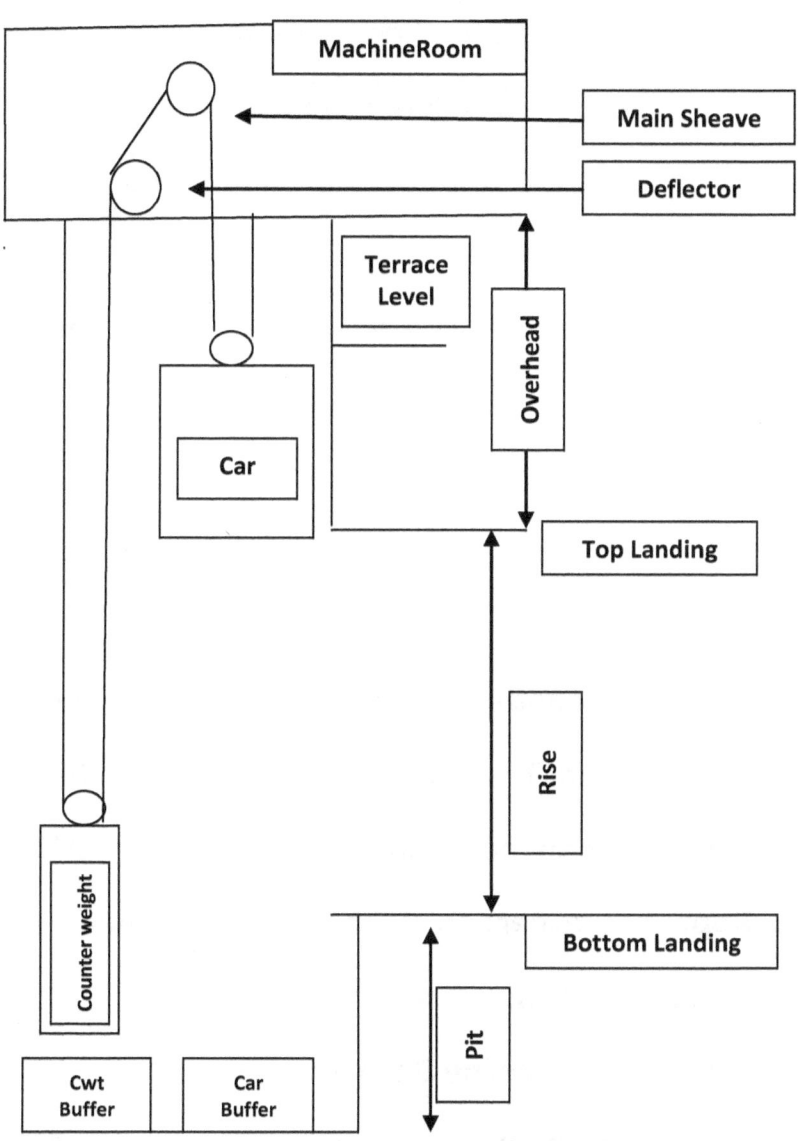

MachineRoom

Main Sheave

Deflector

Terrace
Level

Overhead

Car

Top Landing

Rise

Counter weight

Bottom Landing

Cwt
Buffer

Car
Buffer

Pit

5.7.3 Recent Trends:

One of the leading manufacturers of elevators have introduced Polyurethane Coated Steel Belts in their elevator systems. As per the manufacturer it replaces the heavy woven steel cables that have been the industry standard since 1800. These belts are only 3mm thick and hence make it possible to have smaller sheaves thus making it possible to mount the machine in the hoist way itself thus eliminating the need for machine rooms.

Similarly another manufacturer has introduced "Suspension Traction Media" by which the sheave diameter can be reduced up to 85mm. This Ultra Rope has a carbon fiber core surrounded by high friction coating..

5.8 Driving Machine Sheave:

The machine sheave is just a pulley with grooves around the circumference. The sheave grips the hoist ropes, with car on one side and counterweight on the other side. On account of rope tension, each element of the hoist rope in contact with the driving sheave is pressed against it and becomes a source of friction. When you rotate the sheave, the ropes move along. The friction provided by the groove pressure is known as **available traction.** It is generally accepted that the maximum available traction is dependent upon three major factors:

1. Angle of wrap of the ropes around the traction sheave;

2. Shape of groove profile;

3. Coefficient of friction between rope and sheave material.

Available Traction (TR_{av}) must be greater than the Required Traction (TR_{req}) for the system for the proper operation as described above.

But if the elevator is stalled due to any reason (Safety operation, Counterweight on buffer etc), it must be assured that traction of the system can be overcome so as not to drive the non-stalled component to an unsafe condition. Under such condition, TR_{av} must be less than $TR_{stalled}$.

5.8.1 Groove Types :

The U groove sheave (Fig 2) can be considered as the desired sheave of choice for optimum rope life. The larger D/d ratio makes bending easier, provides a large arc of contact between rope and sheave, reduces operating stresses, and generally optimizes rope life. Unfortunately, in spite of its merits for improving rope life, this type of groove does not provide sufficient traction and hence is used mostly only for high speed applications with double wrap arrangement.

The friction factor F of the ropes in the grooves is dependent on the groove profile.

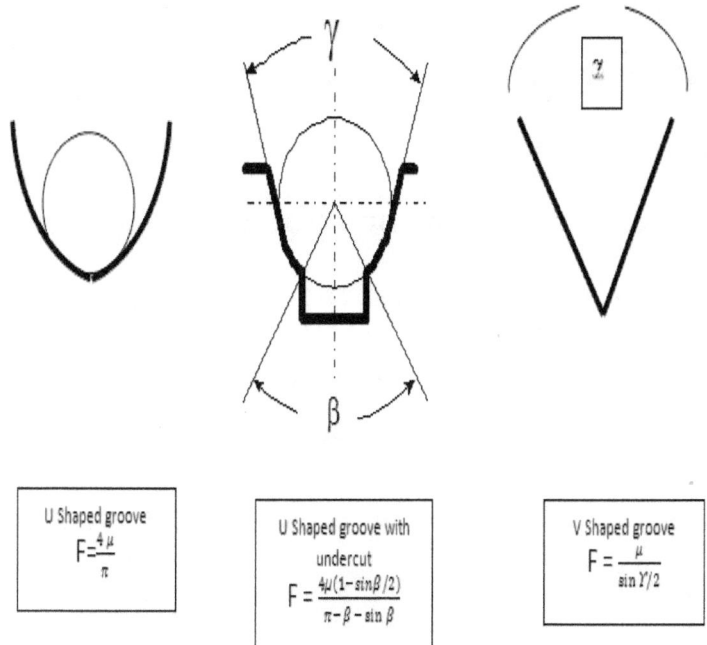

U Shaped groove

$$F = \frac{4\mu}{\pi}$$

U Shaped groove with undercut

$$F = \frac{4\mu(1-sin\beta/2)}{\pi-\beta-sin\,\beta}$$

V Shaped groove

$$F = \frac{\mu}{sin\,Y/2}$$

μ = Coefficient of friction between hoist rope and sheave (Dynamic assumed as 0.09 & Static is

0.2)

γ = Angle of the V groove in radians

β = Angle of undercut in radians

π Radians = 180 deg

1 degree = 0.01746 Radians

173

Refer fig 1 As undercuts become larger, groove pressure increases , traction increases, and unfortunately, rope and sheave wear and rope fatigue accelerates.

Refer fig 3, generally the angle of the V groove is between 32 and 40 deg. Traction increases with reduction in V angle. This type of groove places the largest amount of pressure on the ropes and sheave grooves, resulting in the greatest amount of traction, but also the greatest amount of rope abrasion.

Undercut U- and V-grooved sheaves adversely affect rope performance. But these grooves increase traction and therefore do not require a large diameter sheave, hence these groove types are more preferred by elevator system designers.

5.8.2 Calculation of available traction:

The available traction relation depends on the type of sheave grooving, the arc of contact the rope makes with the driving sheave, and the speed of the

elevator

The available Traction $TR_{av} = e^{f.\alpha}$

Where   e   is the base of natural logarithm

             f   is the friction factor of ropes in the

groove

             α   Angle of ropes on the traction

sheave in radians

             e = 2.718

The friction factor F of the ropes in the grooves is

dependent on the groove profile.

5.8.3 Sample Calculation of Available Traction (Dynamic and Stalled)

Stalled traction is when the counterweight sits

on the buffer whereas Dynamic is at normal condition

of elevator operation.

Groove Type	α	β	γ	Sin β	Sin β/2	Sin γ/2	4μ(1-Sin β/2) Dyn	4μ(1-Sin β/2) Stat	Rad β	(Π-β-Sin β)
U Shaped	135	90		1	0.707	-	0.105	0.23	1.57	0.57
Under Cut $F = \frac{4\mu(1-\sin\beta/2)}{\pi-\beta-\sin\beta}$	135	105	-	0.966	0.793	-	0.075	0.166	1.833	0.341
	180˚	90	-	1	0.707	-	0.105	0.23	1.57	0.57
	180	105		0.966	0.793	-	0.075	0.166	1.833	0.341
.V Shaped $F = \frac{\mu}{\sin\gamma/2}$	135˚	-	35	-	-	0,301				
	180˚	-	35	-	-	0.301				

176

Groove Type	α	β	γ	Dynamic				Stalled		
				μ	F	Rad α	TR$_{av}$	μ	F	TR$_{av}$
U Shaped $F=\frac{4\mu}{\pi}$	135	-	-	0.09	0.115	2.357	1.31	0.2	0.255	1.82
	180			0.09	0.115	3.143	1.42	0.2	0.255	2.22
	360			0.09	0.115	6.286	2.06	0.2	0.255	4.97
U Shaped Under Cut $F = \frac{4\mu(1-\sin\beta/2)}{\pi-\beta-\sin\beta}$	135	90		0.09	0.184	2.357	1.54	0.2	0.42	2.69
	135	105	-	0.09	0.22	2.357	1.68	0.2	0.49	3.17
	180°	90	-	0.09	0.184	3.143	1.78	0.2	0.42	3.74
	180	105		0.09	0.22	3.143	2	0.2	0.49	4.66
V Shaped $F = \frac{\mu}{\sin\gamma/2}$	135°	-	35	0.09	0.3	2.357	2.02	0.2	0.67	4.86
	180°	-	35	0.09	0.3	3.143	2.57	0.2	0.67	8.25

From the above table it is clear that

- Available Traction increases with increased wrap angle

- Traction increases with Increased under cut.

- V cut provides best traction

- Available traction can be increased by increasing the actual coefficient of friction of the material.

Note that all the above parameters are dependent on one another.

Compromising on any of the above factors should not change the final traction value. With this background, elevator system designers need to be very careful in estimating traction and establishing their designs.

5.8.3.1 Required Traction:

When the car and the counterweight hang on the main sheave, the weight on both sides will not be the same. The weight difference varies as the car position changes with respect to the counter weight. The weight of car, counter weight, weight of ropes, weight of compensating cable, weight of traveling cable – all these weights influence the difference in weights on both ends of the sheave. The available traction, which was calculated earlier, must be more than the required traction in order not to allow slip of the ropes and allow the ropes to rotate along with the car. In addition to the weights, in a dynamic situation, the acceleration and deceleration of the car also influence the required traction.

For having proper traction of the system, ,the following equation must be fulfilled.

The Worst Case required traction $TR_{REQ(DYN)} <$ TR_{AV}

The worst case TREQ(DYN) occurs when

- Empty Car is decelerating to stop on top most floor
- 125 % loaded car decelerating to stop at bottom floor.

Hence we need to calculate the traction under the above two conditions to determine the system traction.

For calculation of the traction, we need the following parameters

Car weight = empty car weight + Duty Load

Counterweight = Car weight + OB x Duty + TC

unit weight x Rise /4

ROPE weight = Rope unit weight x # ropes x

(Rise + Overhead) x 2

TC weight = TC unit weight x Rise / 2

Comp weight = Comp unit weight x Rise

Accel=0.5m/s ;

OB (Over balance) = 45%

From the Example 1, we already have the following

elevator specifications

Elevator Specs

Number of ropes	3
Duty Load (Kg) DL	1088
Rise (m)	60
Speed (m/s)	1,5
Gravity (g) (m/s^2)	9.81
Acceleration (m/s^2)	1.0

Dia of Ropes mm	10
Overhead (m)	4.5
Pit Depth(m)	1.6
CWT Over balance	45%
Roping	2:1
Total Car weight (Kg) Carw	1300

From Manufacturer's catalog	
Rope unit weight (Kg/m)	0.35
Trav Cable unit weight (Kg/m) tc	1.1
Comp Rope unit weight (Kg/m) tcomp	2.24
Min braking force of 10mm rope (N)	40000

Calculations:

(Car & CWT sheaves not considered in this calculation)

Traveling cable weight $Trw = R * tc * 0.5$	33
Total Counter weight= $Carw + 0.45 * DL + 0.5 Trw$	1806
Total Comp chain weight $Compw = R * tcomp$	134
Total Rope length= 2* Pit depth + 2* Rise + 4* OH	141.2
Weight of ropes= Rope length* unit weight*No of ropes	148

Condition 1

Empty Car going up stopping at top landing

T1 = (Total mass of Empty Car + Traveling cable + Comp Chain) X (Gravity – Acceleration)

$\quad$ = (1300+33+134) X (9.81-1) = 12924

T2 = Total mass of CWT X (Gravity + Acceleration) + Total mass of Hoist Ropes * Gravity +

$\quad$ Total mass of CWT* Roping^2*Acceleration

$\quad$ = 1806*(9.81+1) + 148*9.81+148*2^2*1= 21566

T2/T1 = $\quad$ 21566 / 12924 = **1.67**

Condition 2

125% loaded car going down stopping at bottom landing

T1 = (Total mass of Empty car + 1.25* Duty Load)*(gravity + Acceleration) + (Total mass of hoist ropes * Gravity) + (Total mass of hoist ropes*Roping^2*Acceleration)

= (1300+1.25*1088)*(9,81+1) + (148 * 9.81) + (148*

2^{2}*1) = 28755+ 1452 + 592 = 30798

T2 = (Total mass of CWT + Weight of Comp Chain) * (

Gravity – Acceleration)

= (1806 + 134) * (9.81-1) = 17091

T1 / T2 = 30798 / 17091 = 1.8

Greater of the above two ratios is 1.8Available

Traction with U undercut 105° and wrap angle of

180°is 2 .

Hence TR$_{AV}$ >   **TR$_{Req(Dyn)}$**   and   meets   the

requirement

Traction under Stalled condition:

It is also important to ensure that when the

Counterweight is sitting on the buffer, the available

traction must be less than the required traction so

that the traction should not be large enough to lift the

car.   **TR$_{REQ(STALL)}$>   TR$_{AV(STALL)}$**

T1 Stall = Total mass of (Empty car + Traveling Cable + Comp Chain) * Gravity

$$= (1300 + 33 + 134) * 9.81 = 14391$$

T2 Stall = Total Mass of Rope * Gravity

$$= 148 * 9.81 = 1452$$

$$TR_{REQ(Stall)} = T1_{STALL} / T2_{STALL} = 14391/1452 = 9.9$$

$T_{AV(STALL)}$ for U sheave with 105° undercut and 180° wrap angle is **4.66** Hence $TR_{REQ(STALL)} > TR_{AV(STALL)}$

CHAPTER 6

RAILS, GOVERNOR, SAFETY, BUFFERS,

BRAKES, LIMIT SWITCHES

6.1   RAILS :

Both the car and the counterweight must be guided by at least two rigid steel guide rails. The functions of guide rails are as follows:

(1) to guide the car and the counterweight in their vertical travel and to minimize their horizontal movement,

(2) to prevent tilting of the car due to eccentric load, and

(3) to stop and hold the car on the application of the safety gear.

The guide rail is considered a continuous beam with variable supports.

As per Indian standard, the Guide Rails shall be of " T " section. The strength of the guides, their attachments and joints shall be sufficient to withstand the forces imposed due to the operation of the safety gear and deflections imposed due to uneven loading of the car.

In the calculation of the guide rails, two operating conditions should be taken into consideration (1) safety gear operation, (2) running conditions with the load unevenly distributed on the car floor. The buckling stress has to be calculated as per guidelines by Indian Standard.

Guide rails are typically, 5 meter (16-feet) long "T" – sections manufactured in different sizes for different duty loads, speeds and classes. For rail end-to-end connections, plates are used to hold and these are called fishplates. The Rails and are attached to the building with rail brackets. These brackets known as rail brackets keep the rails plumb, square and maintain the correct distance between the guide rails (DBG) so that the car maintains proper ride quality.

Guide rails are a significant factor in achieving ride quality. The characteristics of a rail which impact the ride quality includes blade straightness, rail height, tongue and groove width, blade thickness, surface finish, tongue and groove positioning etc

Given proper machining and accessories, the installation of the rail system is the critical and final stage of the process of making a good set of rails. The better rail systems are those requiring the least amount of time and effort during field installation. The quality of an installation is dependent on the alignment of the rails, both horizontally and vertically.

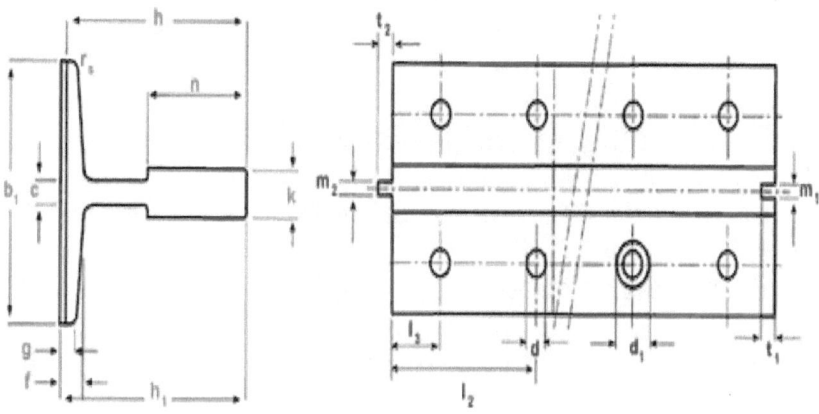

Typical Guide Rail with markings of dimensional details – Source: Savera

Rails are lubricated with oil to reduce the frictional resistance and enhance the riding smoothness. The present day shoes have built in lubricator tanks with feeding wicks through which oil trickles to the sliding surface of the guides. Oil or grease is likely to obsorb dust and also changes its properties due to temperature and other environmental conditions.

6.2 SAFETY GEAR & GOVERNOR:

The Governor monitors the movement of the cab and safely stops the elevator if the elevator car over speeds. The over speeding could be due to malfunction of speed control of the elevator motor or due to free fall of the cab due to breakage of elevator ropes. The Governor activates an over-speed switch if the cab moves greater than the rated speed. The switches are used to stop the elevator and apply the brake.

The worst case of over speeding of the cab is when the traction ropes breaks and when the cab falls down due to gravity. Under such worst case conditions, the governor, which is located in the machine room or

GOVERNOR

Overhead depending on the elevator design activates the safety block mounted on the car frame and the safety block stops the elevator movement by locking on to the elevator rails.

The governor rope runs over the governor sheave and down to the elevator car and is attached to the safety trip mechanism. The governor rope continues to the pit and runs over a pit mounted sheave and turns back to the governor sheave located in the machine room or overhead. This governor rope arrangement forms a continuous loop while the elevator moves up and **down the hoistway.**

If an elevator over speeds either because of a major malfunction or the suspension ropes are severed then the governor sheave accelerates as the car accelerates until a preset speed limit is attained and a speed sensing device in the governor is tripped. At that point the governor's rope griping jaw is activated to seize the governor rope. As the car continues to descend, the stopped governor rope moves the fixed car safety operating lever to engage the safety jaws installed at the ends of the safety plank. Friction between the safety jaws and the guide rails slow the car to a halt.

The safety jaws are of two types, instantaneous and progressive. Instantaneous Safety Gear is a device in which gripping action on the guide is almost immediate and the cab almost stops instantly. Instantaneous Safety is normally used below a speed of 1 mps (meter/second)

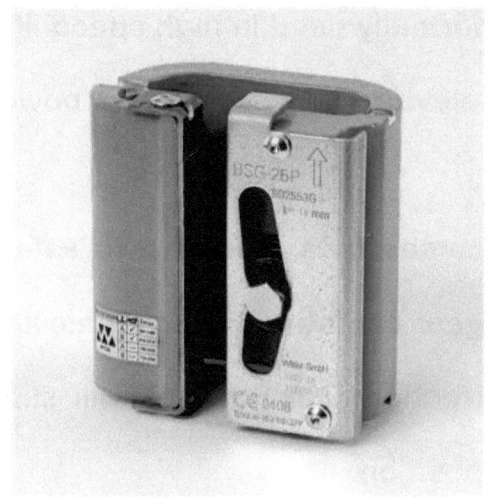

Progressive Safety Block

A progressive gear is a safety in which deceleration is effected by a braking action on the guides and progressively brings the car to a stop within the limits prescribed by the standards. Progressive safety is normally used for higher speed lifts as a sudden stop may cause severe jerk and may injure the passengers inside the cab.

The governor for few high speed designs, are encoder capable. Mounted on the governor, the encoder tracks the precise location of the elevator and relays the information back to the controller.

This arrangement is normally used in high speed lifts and to ascertain the elevator position after a power failure.

Once the car comes to a stop due to safety operation, the safety gear can be released by moving the car in upward direction. The wedges will slide back to their original position.

6.3 BUFFERS:

Buffers are located in the pit as a final safety device to bring the descending car or counterweight to rest on it if they move past the normal downward limit of travel. The buffers are designed to absorb the kinetic energy of motion of the car.

Buffers are used both on the Car as well as Counterweights to absorb the energy either in UP or DOWN travel.

Buffers are of two types: Spring and Oil buffers. Spring buffers can be used for elevators with lower speeds and Oil buffers for higher speeds.

Spring Buffer

Oil Buffer

As a thumb rule the buffer should be adequate to support a static load of 2.5 to 4 times the mass of the car together with the passenger load without compressing the buffer solid.

6.4 BRAKE:

In lift application the brake is normally electromagnetic type. The major versions in use are Shoe type and Disc type. The brake is applied by spring force in normal condition. It is released by energizing the coil of the electromagnet which in turn will release the drum or disc allowing it to rotate. The electromagnetic coil is designed to overcome the spring force. In the event of power failure the brake automatically applies due to spring force.

As per standard the brake should have sufficient torque to safely stop a car with 125% overloaded also hold the loa of the system.d at stop. Static torque is to hold the load at rest and dynamic torque to absorb kinetic energy of all moving parts.

The design of the magnet coil is normally done with DC voltage. In AC coils the operation will be abrupt while closing or opening which produces heavy jerk. DC coil operation is much smoother and softer. Normally brake coils are provided with levers for manual release in case of emergency.

Lifts are equipped with various controls to slow down and practically stop the lift dynamically without the mechanical brake coming into opration.Function of the brake is to hold the car in its position after it comes to level and stop.In case of single speed lifts with no speed control, the lift is stopped purely by mechanical control resulting in poor leveling accuracy and frequent wear and tear of brake liners

6.5   LIMIT SWITCHES:

These are limit switches mounted in the shaft and operated by the movement of the car. Generally minimum of 3 limit switches are mounted at the top and bottom floor regions. The first switch encountered near the topmost or bottom most floor is the terminal slow down switch. During normal conditions, the floor positions are counted by the controller with the help of the photo electric switch mounted on the car top and initiate slow down at the respective landings. In the event of failure of this counting, the terminal slow down switches initiate slow down of elevator to prevent over travel. The second set of limit switches apply the brakes in the event of failure of counting of pulses by the processor and also the failure to slow down with help of terminal slow down switch. If the car approach the terminal landing at normal speed without slowing down the second set of limit switches operate to stop the car safely. The third set of limit switches operate when the car over travels by about 50 to 60mm beyond the terminal floor. These are known as final limit switches. These switchers disconnect the power to the controller and drive and the lift can be restarted only if the car is physically moved to disengage the final limit switches. All these switches ensure that the car does not over travel But in the event the car over travels the car or counter weight, depending on the direction of travel, sit on the buffer and the kinetic energy is absorbed by the buffer. The buffer also has a switch which once again trips the elevator motion.

The other safety switches are:

(i) Stop button inside the car. This button was available in the car operating panel and the car can be brought to an abrupt strop by pressing this button. This button causes more inconvenience such as stopping with heavy jerk, stopping out of floor level and passenger entrapment. In the recent designs, this switch has been removed from the car operating panel except in certain states like west Bengal where it is a code requirement.

(ii) Over speed governor switch: This switch is located in the governor and trips and brings the car to a halt when the car speed exceeds about 1.15 times the rated speed. This type of over speed may be due to malfunction of controller or the drive. The Governor trips the safety block if the speed exceeds

1.4 times the rated speed. This kind of excessive speed may be due to rope breakage and free fall of elevator.

(iii) Safety Gear Switch: Whenever the safety gear is tripped due to governor over speed detection, the safety gear, in addition to operating the mechanical safety, also switches an electrical safety to cut off the lift movement by the controller.

(iv) Stop switch on the car top: This switch is normally used by the service mechanic when he travels on the car top.

(v) Stop switches in the Pit: Normally two switches are provide, one inside the hoistway very close to the sill and accessible from the landing. The other switch is located on the walls of the pit. The service person, while entering the pit puts off the switch near the sill and after entering the pit switches off the other switch mounted on the wall of the pit also. While coming out of the pit, he resets the switch inside the pit first and then the one near the landing sill after coming out of the hoistway.

All these safety switches are connected in series and will be connected suitably to stop the motion of the lift.

Yet another emergency device is the door open key. As we know, all doors in closed condition and mechanically locked is the basic safety requirement. In order to rescue the trapped passenger, it may be required to open the respective landing door near which the car has stopped. For this purpose, special keys are provided by the manufacturer and kept in a secured place. This an emergency safety device and must be used only by authorized personnel and well trained in rescue operation. This is a very difficult operation as many accidents occur while doing this kind of emergency operation.

6.6 Emergency devices:

In elevators there is a risk of passenger getting trapped inside a car either due to power failure or due to failure of the lift control system. A battery operated alarm button is provided in the car operating panel to sound an alarm in order to get external help to rescue the stranded passenger. Hands free interphones are also provided to talk to the security or any other assigned person in the building.

6.7 Requirements for fireman's lifts:

Fireman operation helps the Fireman to use the lift and conduct rescue operation when the building is on fire.

a) For buildings having height of 15m or more, at least one lift will meet the requirements of fireman lift.

b) The lift assigned for fireman operation must have minimum of 1.44 sqm of floor area.

c) The capacity of the lift has to be atleast 8P (544 Kg)

d) Doors shall have automatic operation (both car and landing) with a minimum of 800cm opening.

e) Landing doors shall have minimum of 1 hour fire resistance.

f) The speed of the elevator should be such that the lift travels from bottom to top floor within 1 minute.

CHAPTER 7

AUTOMTIC RESCUE DEVICE AND

OPPORTUNITY FOR DEVELOPMENT

7. Automatic Rescue Device (ARD):

In countries such as India where power interruption is frequent, Automatic Rescue Device is widely used to rescue the passengers trapped inside the lift cabin.

7.1 Manual Rescue:

In the past, when Automatic rescue devices were not available, elevator mechanics used to lift the machine brake and rotate the fly wheel to bring the lift to the door level and then open the door using a key to free the passengers. This method is highly unsafe and was required to be conducted only by trained personnel. Remaining inside the lift cabin till the arrival of trained personnel caused lot of anxiety and panic for the trapped passengers. Lift industry has witnessed many incidents during and before rescue operations.

7.2 Stopping Jerk:

Referring to para 2.6.3.2 in this book, apart from passenger entrapment , when the elevator stops suddenly due to power outage, the PM Gearless elevators seem to introduce heavy jerk as compared to equivalent geared elevators. This causes discomfort especially to elderly and pregnant women. This severe sudden stopping jerk is in addition to the passenger entrapment and anxiety till the person comes out of the lift. Typically most of MRL lifts are gearless.

At present when the book is written, many lift manufacturers have improved the engineering design to reduce the stopping jerk on power failure.

7.3 Automatic Rescue Device Basic Operation:

At present, for rescuing a trapped passenger, automatic rescue devices are used which is a safe way of rescuing the passengers. ARD is s control device meant to bring a stuck elevator between floors, due to loss of powe,r to the nearest level and open the doors in order to allow the trapped passengers to be evacuated. Such a device use auxiliary power such as batteries and it must comply with all the safety requirements of the lift as during normal run. In the case of manual door operation, the device shall allow the passengers to open the door on reaching the level and in case of power operated doors; the doors will automatically open on reaching the floor level. The speed during the ARD run is usually less and typically it is about 1/10 to 1/16 of the rated speed. Since the ARDs take over automatically, the passengers are rescued shortly after power failure. But the ARDs are useful only for entrapment due to power failure and not when the elevator stops due to failure of controller or any of the safety switches.

7.3.1 ARD Building Blocks:

There may be many variations between each manufacturer, but typically following are the major parts of an ARD system:

7.3.1.1 Battery:

Generally 4 Nos of 12V Lead Acid Maintenance free batteries are used to power the elevator during rescue operation. The AH of the battery is chosen based on the power requirements of the elevator, typical ampere hour rating being 7.2AH, 18AH or 24AH. .

7.3.1.2 Charger and Charger Transformer:

During normal power, the batteries are charged by the Charger transformer and the charger. The charger transformer steps down the three phase 400V AC mains power and feeds to the charger.

7.3.1.3 Power Board and Power Transformer:

The power board converts the 48V DC battery supply to suitable three phase AC and the power transformer steps up the output to feed 400V AC 3 phase to the elevator. The power board consists of IGBTs or power MOSFETs which are driven by drivers controlled by PWM circuitry.

7.3.1.4 Logic Board:

The logic board typically does the following functions:

a) During normal power, it switches main contactor to provide 400V AC 3 Phase 50 HZ mains power to the elevator . On loss of mains power, it switches off the main contractor and switches ON the ARD contactor to provide

400V AC 3 Ph 50 HZ emergency power to the Elevator.

b) Detects the power failure signal from other external device such as single phase preventer and initiates ARD operation..

c) Provide signal to the elevator controller / Main VVVF drive to indicate that the power available is emergency power.

d) Provide signal to the voice annunciator to make emergency rescue announcement inside the car.

e) Detects Door Zone signal from the elevator, terminates the ARD run.

f) Continues and finishes Rescue operation even if mains power resumes in between.

g) Detects the current drawn from the battery and provide command to the main drive to reverse the direction of travel in order to make the movement of elevator in the direction of gravity.

h) Detect the health of the battery and provide prior indication during normal power.

i) Has many timers

(I) 6 sec (approx.) timer to initiate ARD operation

(ii) Initiate another timer typically 3 min to abort ARD operation if door zone signal was not received.

(iii) About 15 sec after receipt of Door Zone signal to allow the elevator controller to fully open the door and then completes ARD operation.

j) Disconnect s the battery from discharge inside the
ARD circuitry if power is shut down for a prolonged time.

7.4 Advancement in main drive technology:

During rescue operation, first, the ARD tries to run the elevator in a pre-determined direction. If the battery current drawn is more, the elevator run is stopped and direction of travel reversed to take advantage of gravity. This feature drains the battery and also takes a little longer for rescuing the passenger. In the current designs, the main VVVF drive has been designed to remember the running current and direction of travel during each run. On power failure, during ARD run, the drive runs in the easy direction thus avoiding unnecessary current drain from the battery.

7.5 Seamless ARD run:

Prevention of "stopping jerk" due to power failure and subsequent ARD run can be overcome if the elevator is run on 3 phase un- interrupted power supply (UPS). Though few customers use UPS, it is an expensive solution..

7.5.1 GeN2 Switch by OTIS:

Switch works on single phase supply with the regenerative drive powered by lead acid battery. But at present, this is suitable only up to 8P, 1 mps MRL elevator.

7.5.2 TOSMOVE NEO by Toshiba:

Uses Super Charge ion Batteries (SCiB). When power failure occurs, gradually reduces elevator's speed, and move to the nearest floor at slow speed. During continued power failure, moves at 1 mps for 30 minutes.

7.5.3 Affordable seamless ARD

Similar to OTIS and Toshiba, Schindler has a product which works on solar power and single phase supply. But in the opinion of the author, these products are not able to provide a low cost solution to the average

Indian buyer. The author has also tried few options of providing seamless ARD run. But the author's efforts were also in the high cost category. Basic block diagram of the seamless ARD as designed by the author is given below.

7.5.3.1 Description:

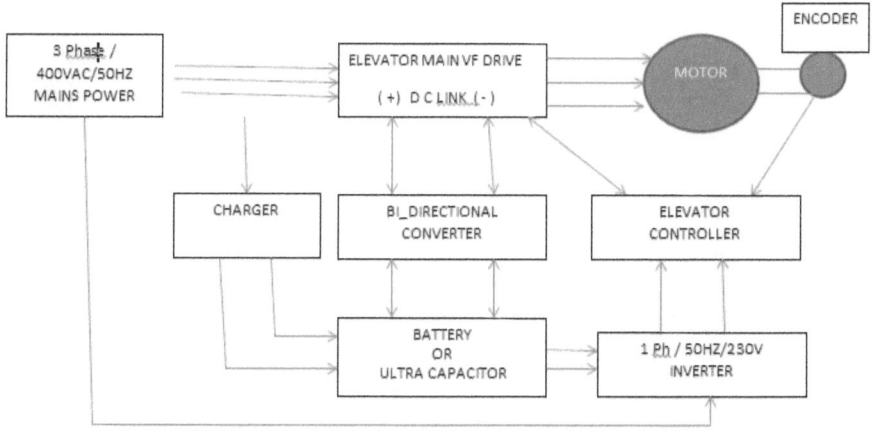

During normal mains power, 3 Phase mains power charges the battery . The main VF drive and the Controller power is provided by the mains. During elevator run, the battery is also getting charged from the DC link bus of the drive. The bi directional converter transfers energy from the DC link to the battery / Super capacitor. In the event of power failure, the DC link is fed by the battery through the Bi directional bus. Supply to the controller is maintained by the DC to AC inverter.

Above is mentioned only as an idea for future work as the solution tried was not fully tested at site and or commercialized.

ANNEXURE – A

STANDARDS AND CODE

Standard is a level of quality that people expect and generally accept as normal. Elevator standards are a set of rules and norms to be observed right from the design stage, manufacturing, installation and maintenance. Almost every country has evolved its own standards to suit their local conditions. In India we have the standard formulated by the Bureau of Indian standards, BIS, and its relevant number for Lifts is BIS 14695. This standard periodically gets updated by a technical committee comprising of experts in the industry derived from general public, manufacturing companies, town planners, government agencies etc.

By and large, there standards are similar worldwide except for few variations to suit the respective countries.

Standards give brief description of functioning of mechanical/ electrical/ electronic components and assemblies. Safety aspects are given high emphasis.

The main idea of standards is to have a uniform code for lifts all over the country .Many of the states in the country have their own set of rules and codes. By and large these state level rules and codes prescribe norms and procedure for obtaining permission to install a lift and thereafter obtain the license for operating the lifts. There are moves in the states to follow the Indian codes and for the Indian code to align with international codes.

By referring to the standards the architect / builder can obtain most of the required details on the dimensions & sizes based on the types of lifts planned their capacity, speed etc. Sometimes one may not be able to follow the guidelines due to other space constraints. Drawings are finalized after mutual discussions and agreements between the supplier and purchaser. Additional care is taken not to dilute the safety aspects of the code.

REFERENCES

1) Lubomir Janovský. <u>Elevator Mechanical Design: principles and concepts</u>.

   England: Ellis Horwood Limited, 1987.

2) George R. Srakosch. <u>The Vertical Transportation Handbook Third Edition.</u>John Wiley & Sons, Inc., 1998.

3) The Vertical Transportation Handbook edited by George R. Strakosch, Robert S. Caporale

4) Practical Variable Speed Drives and Power Electronics by Malcolm Barnes

5) Mechanical and Electrical Equipment for buildings by Walter T, Grondzik, Alison G Kwok, Ien Stein, John S Reynolds

6) Generation and Utilization of Electrical Energy by S Sivanagaraju.

7) Lift Technology by K Subramaniam.

8) OTIS by Jason Goodwin

9) Indian Standard BIS 14665

10)   http://www.mitsubishielectric.com/elevator/overview/elevators/history.html)

11)   http://www.ushamartin.com/wp-content/uploads/2014/07/Elevator-Rope-Catalogue.pdf

12)   www.**bridon**.com/usa/x/downloads/usatechnical/**traction**.pdf

13)   http://web.mit.edu/2.972/www/reports/elevator/elevator.html

14)   Elevator world – Continuing technology

15)   www.otis.com

16)   www.schindler.com

17)   www.wwwrope.com/